CONTENTS

LCCN: 87-070256
ISBN: 0-916150-84-4

"In this village, John Froehlich built the first gasoline tractor that propelled itself backward as well as forward. Far-reaching in its effect on modern agricultural history, it moved out of this village and into the world in 1892."

So reads the first paragraph of the inscription on a monument in Froehlich, Iowa, a town named for Froehlich's father. A U.S. Department of Agriculture publication provides a few more details:

"...John Froehlich mounted a gasoline engine, built by the Van Duzen Gas and Gasoline Engine Co. of Cincinnati, on a Robinson running gear equipped with a traction arrangement of his own manufacture. This was probably the first gasoline tractor of record that was an operating success. It completed a 50-day threshing run...pulled the thresher over difficult terrain and operated in temperatures of – 3° to 100°F."

Operating success? Perhaps. That threshing run saw over 72,000 bushels of small grain threshed.

But commercial success? Hardly. Only two Froehlich tractors were ever built. Both were sold, and both were returned by dissatisfied buyers.

Nevertheless, that threshing run had generated enthusiasm in the hearts of Froehlich and his friends. That enthusiasm produced action. The second paragraph of the inscription on the monument reads:

"Later that year (1892), Mr. Froehlich joined with others in organizing the Waterloo Gasoline Traction Engine Company which later became the John Deere Tractor Company."

The period embraced by that second paragraph was 26 years; the first 20 filled with frustration for the tractor makers. There were the two tractors that were sold and returned. In 1896 a new model was introduced; but only one was sold. Another model was introduced in 1897; but again only one such tractor was sold.

In the meantime, the company turned to the manufacture of stationary engines to produce some income. In 1895 the company was reorganized. Perhaps symbolically, the word "Traction" was dropped from the new company's name and Froehlich, whose interest was tractors, withdrew.

Still, tractor experimentation continued. In 1912 the Waterloo Gasoline Engine Company introduced the first "Waterloo Boy". In 1914, for the 1915 model year, came the Model "R" and two years later the "N" Waterloo Boy. Like the "R," the "N" burned kerosene. And...it had two forward speeds as compared with the "R"'s one; it produced 12 hp at the drawbar, 25 hp on the belt pulley. It also produced sales; by 1918 more than 8,000 Waterloo Boys had been sold and were at work on American farms.

The "N" was in production when, as that inscription on the monument says, the company "...became the John Deere Tractor Company."

In Moline, Illinois, Deere & Company had experimented with tractor design and manufacture. A pilot run of one model had been produced and sold, notably in the Dakotas. With the acquisition of the Waterloo factory, those efforts ended. Tractor development was concentrated at Waterloo.

Traditions of invention, innovation, improvement, and growth had characterized John Deere from the company's beginnings in the mid-1800s. Now those traditions found a new medium of expression—tractors. John Deere engineers merged their talents with those of the Waterloo tractor makers—specialists in engines, chassis, transmissions and power trains—to design and build even better tractors. Continually better tractors.

They are still doing that today.

On the following pages you'll read about the evolution of John Deere Tractors...the venerable Model "D", first Waterloo tractor to bear the John Deere name; the "GP", first with four sources of power; and more, much more.

We wish you happy, nostalgic reading. And as you read, please keep in mind the worldwide nature of agriculture today. This booklet identifies John Deere Tractors as they've been sold in the United States. Sometimes in Canada, more often in other parts of the world, John Deere Tractors have carried different model numbers.

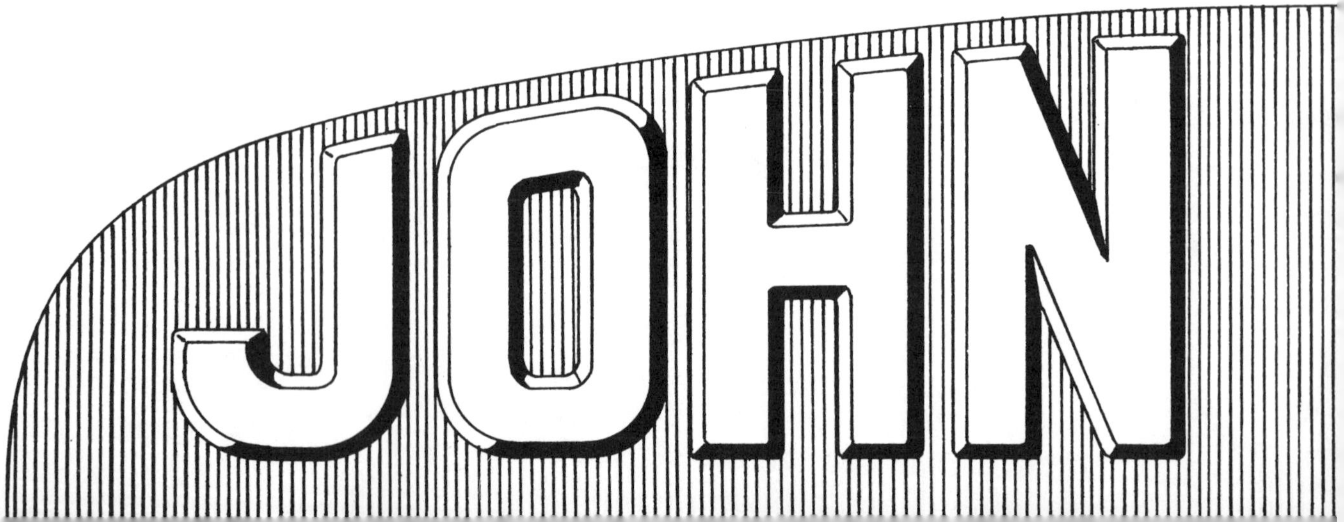

The Froehlich Tractor—1892

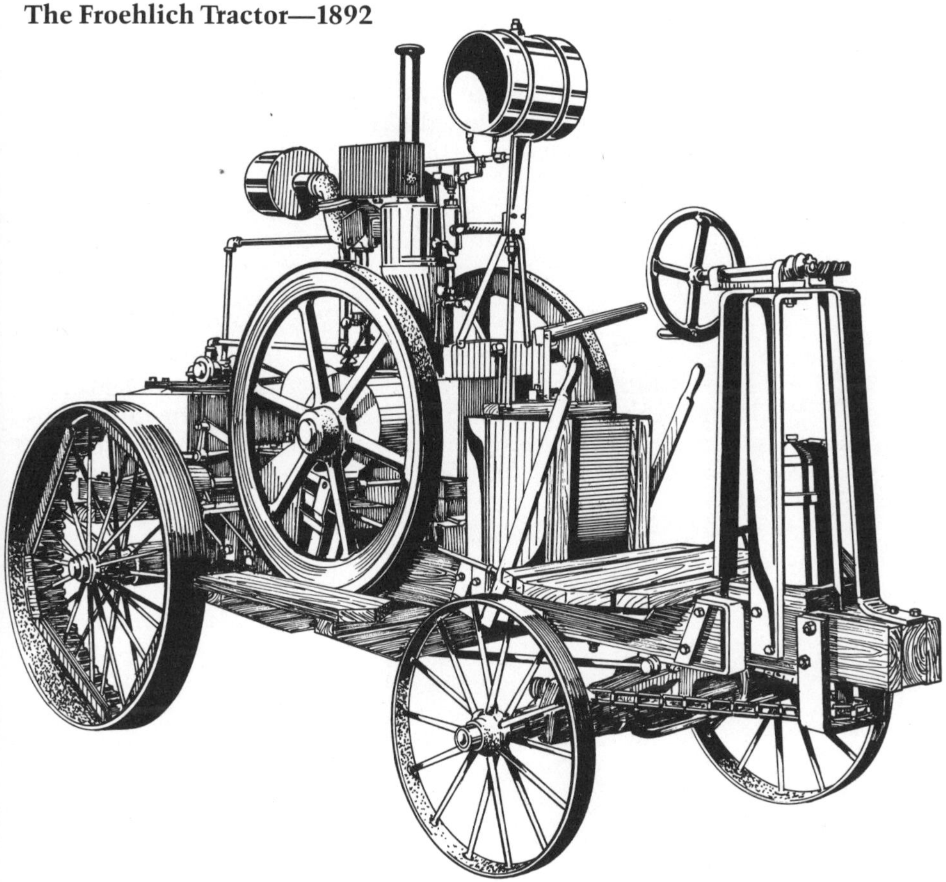

Progenitors of John Deere tractors: the Froehlich (left), first gasoline tractor to propel itself backward as well as forward, and the Waterloo Boy Model "N" (below). Introduced in 1917, the "N" was in production when the company John Froehlich helped organize became the John Deere Tractor Company in 1918. The "N" remained in production another six years until it was replaced by the "D".

The Waterloo Boy Model "N" produced 12 drawbar and 25 belt hp with its 2-cylinder, kerosene-burning engine operating at 750 rpm. Its features included a combined pressure feed and splash lubrication system; pump, fan, and radiator cooling system; cone clutch; open spur gear and pinion final drives. Forward speeds: 2¼ and 3 mph. When Nebraska initiated its tractor testing program, a Waterloo Boy Model "N" was the first tractor tested. Dates of that test: March 31 to April 9, 1920.

Model "N" Waterloo Boy—1917-1924

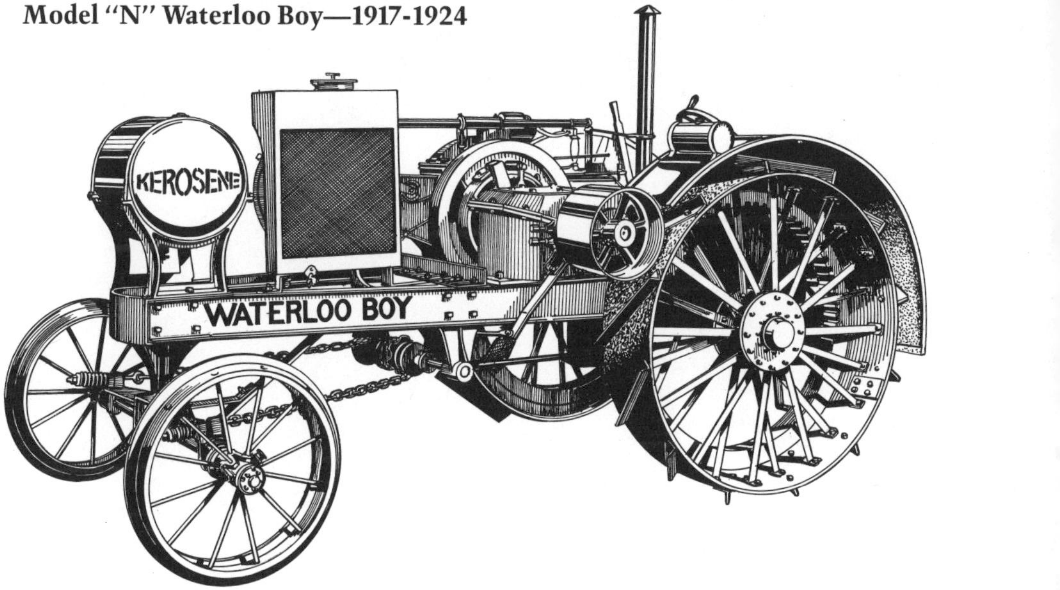

"D"—1924-1953

The "D" was the first Waterloo tractor to bear the John Deere name. And, the most venerable, staying in the line from its introduction in 1923 until the last "D" rolled off the assembly line and into history three decades later, in 1953.

Many factors contributed to the "D"'s popularity:

It operated on almost any low-cost fuel, owners frequently claiming it would burn "anything that could be poured in its tank."

Parts were simple and easy to reach; repairs could be made standing up. Farmers, most of whom made their own repairs, appreciated that.

A low initial cost put the "D" within reach of many farmers who were strapped by the post-World War I depression.

It could pull three 14-inch bottoms in most soils.

Finally, John Deere had established excellent dealer and service organizations, providing easy access to needed repair parts.

The "D", like all succeeding John Deere Tractors until 1960, had a rugged, dependable 2-cylinder engine (you can read more about it on page 9). That engine plus big, rugged gears; one-piece transmission case; and pressure lubrication to main and connecting-rod bearings and piston pins made the "D" an extremely durable machine. Many are still on the job today.

Over the years, of course, the "D" was improved with rubber tires, electric starter and lights, a powershaft, and other features as those innovations were incorporated in other John Deere Tractors.

The "D" was originally rated at 27 belt-pulley hp and 15 drawbar hp. See State of Nebraska Tests No. 102 (1924), No. 146 (1926), No. 236 (1935), and No. 350 (1940).

Note: See page 68 for list of Nebraska Tests of John Deere Tractors.

"GP"—1928-1935

By the mid-1920s, tractors had proved their ability to pull not only plows but other seedbed preparation equipment as well. They were firmly established as the primary source of farm power for the future. Now the engineers began exploring new uses for tractor power. By 1928, John Deere was ready to introduce a more versatile tractor capable of a wide variety of farm work. Its name: "GP".

The initials stood for "General Purpose" and no phrase could better describe its significance. This was the first tractor with four sources of power—drawbar for pulling, belt pulley to drive small shellers or hay presses, power take-off for mowers or small combines, and (an industry first) power lift to raise or lower such integral equipment as planters or cultivators.

The "GP" had an arched front axle and high clearance so it could straddle one row and cultivate as many as three. Naturally, a line of John Deere 3-row equipment was available for use with the "GP".

The first John Deere Orchard Tractor was a variation of the "GP". The "GP" Orchard model had deep-skirt fenders covering the rear wheels down well below the hubs. This prevented limbs from catching in the spokes. The fenders also covered the belt pulley on one side and flywheel on the other, shielding those projections which could catch and bruise tree limbs.

The "GP" had a 2-cylinder, kerosene-burning engine rated at 10 drawbar and 20 belt-pulley hp. Modifications in 1931 raised these ratings to 15.52 and 24.30 respectively. See State of Nebraska Tests No. 153 (1928) and No. 190 (1931).

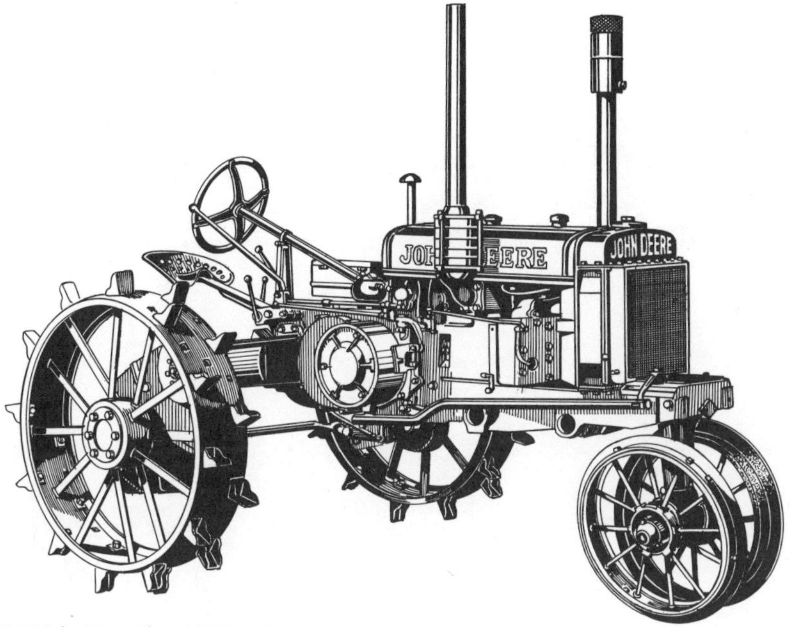

"GP" Wide-Tread—1929-1933

Farmers welcomed the power lift on the "GP", with its promise to ease the manual labor required to raise and lower integrally mounted equipment. And the Corn Belt accepted the 3-row concept. But cotton growers preferred 2- and 4-row operations. Obviously, the 3-row "GP" would have to give way to a new approach.

It did... and soon. In 1929 came the "GP" Wide-Tread (above), identical to its predecessor except that it had a long rear axle, to straddle two rows, and the first John Deere tricycle front end to fit neatly between the two rows the rear wheels straddled. More specialized "GP" tractors followed, including a special model for potato growers.

Finally, in 1932, still another "GP" Wide-Tread was introduced. This model had a tapered hood to provide visibility that the literature of the day described as "... fully equal to that of any horse-drawn cultivator." Steering was improved because front wheel axles were attached to the spindle mount an inch to the rear of the center. Thus the spindle pulled the front wheels caster fashion, preventing whipping, particularly on rough ground. For the operator's comfort, this model had the first John Deere tractor seat mounted on adjustable cushion springs. Individual rear wheel brakes assisted turning. Controls were relocated, with spark and throttle levers mounted on the steering wheel support so the operator could change speed without changing his position or moving his vision from the work area ahead.

Of all the "GP" innovations, power lift was the most popular—and the most quickly copied by other tractor manufacturers.

For power ratings and Nebraska test numbers, see "GP" (left).

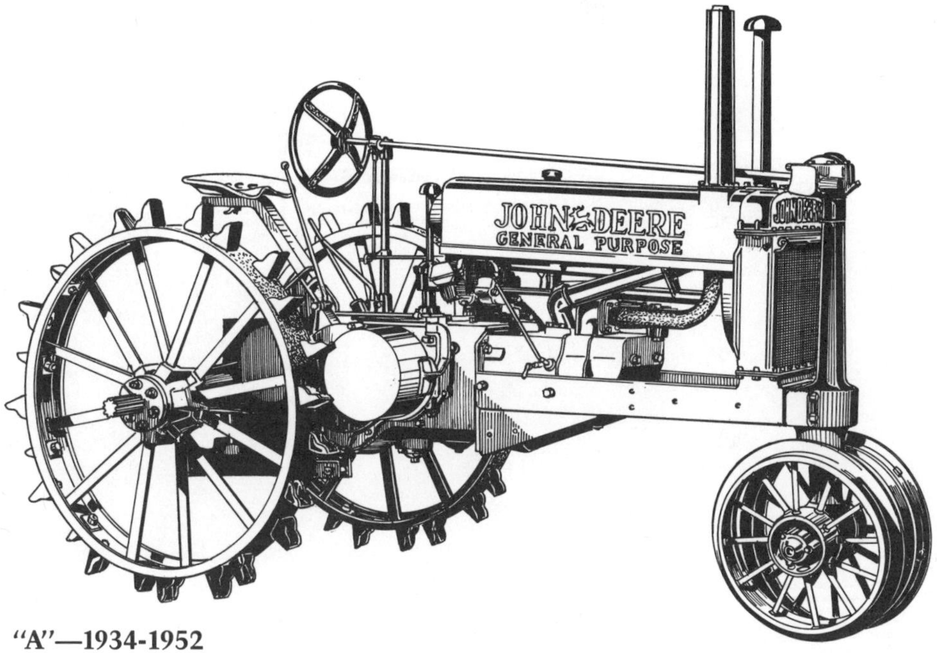

"A"—1934-1952

How wide is a row? 42 inches. Why? That's the average width of a horse. Historically, row width was determined by how close the horse could be driven to the previous row. Early multi-row equipment adopted the 42-inch standard. Early tractors, built to pull that equipment, froze that standard into steel with drop housings and fixed tread.

But problems arose: a fixed tread imposed limitations on farmers beyond certain specific row spacings. It limited the attaching points available to equipment designers. Plows, which worked to the right of tractors, developed side draft.

In 1934, John Deere introduced the Model "A" with answers to all these problems. Two industry firsts, adjustable wheel tread and a one-piece transmission case providing high under-axle clearance, solved the first two problems. And the implement hitch, as well as the powershaft, was located on the centerline of the tractor, effectively reducing side draft.

Other improvements on the "A" included:

Power lift converted from a mechanical to a hydraulic system, increasing both efficiency and speed of operation. It also provided a "cushioned" drop of equipment.

Differential brakes geared directly to the large drive gears, increasing mechanical efficiency and easing operation.

Farmers welcomed the "A" and its variations so strongly that this became the most popular tractor in the company's history.

The original "A" was rated at 16.22 drawbar hp and 23.52 belt-pulley hp. See State of Nebraska Tests No. 222 (1934), No. 335 (1939), No. 378 (1941), No. 384 (1947) and No. 429 (1949).

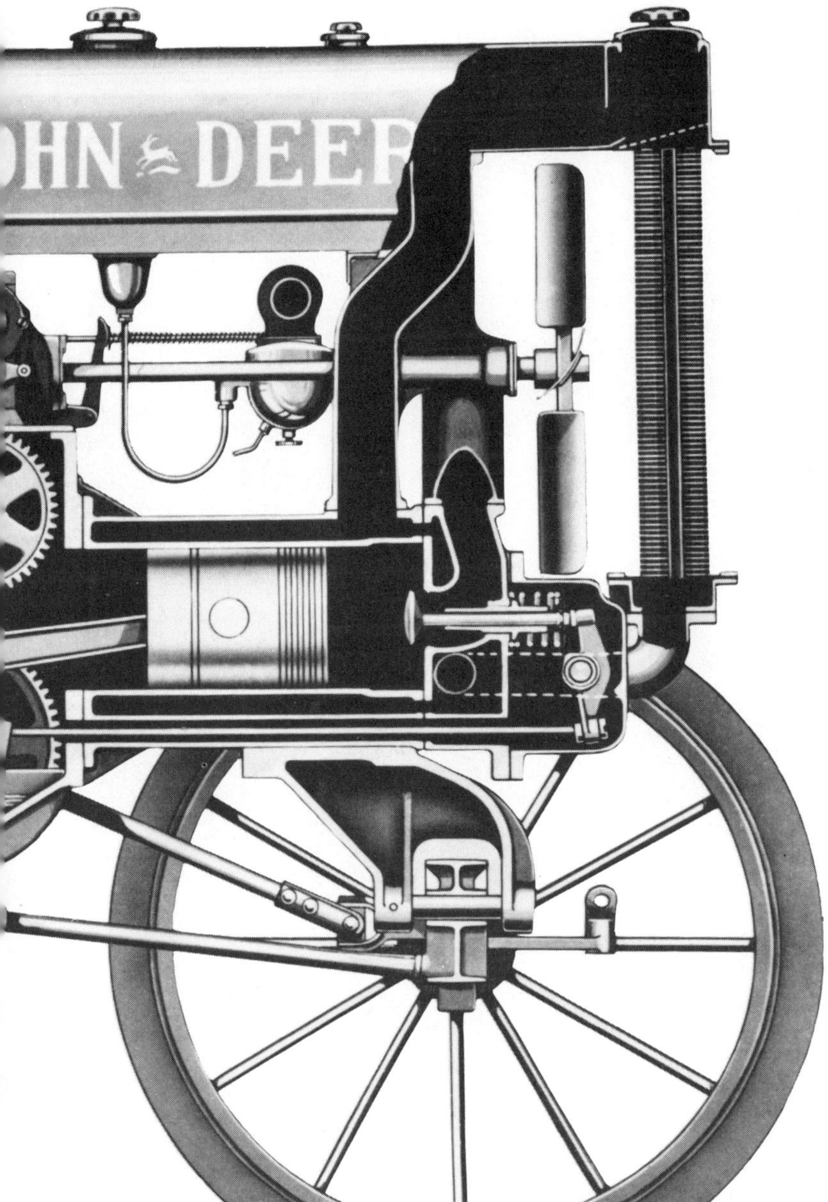

From 1918 until 1960, from the "Waterloo Boy" to the "New Generation of Power", through the evolution from animal horsepower to tractor horsepower, all John Deere Waterloo Tractors shared three characteristics—the name; green and yellow paint; and 2-cylinder, horizontal engines.

This was the four-decades-long era of the tractors called "2-lungers", "Poppin' Johnnies", and a multitude of other nicknames born of affection and admiration. When that era ended, John Deere received rivers of tearful mail. Great men and common, inside the company and out, shuddered to think of John Deere Tractors without the honored 2-cylinder engines. Some expressed fears for the continued existence of the company.

What was so remarkable about the 2-cylinder engine? Everything!

Two cylinders meant fewer moving parts than in engines with more cylinders. This meant fewer parts to wear, to require adjustment, repair, and eventual replacement. Each part could be larger, heavier, more rugged, able to stand up under long hours at full or three-quarter load. All of this translated into stamina, and it helped establish the John Deere reputation for dependability.

Simplicity also meant the engine was easier to understand, easier for a farmer-owner to service and repair himself. Parts were easily accessible and easy to handle because of their larger size.

The 2-cylinder engine was compact, mounted so the crankshaft was crosswise to the chassis (in fact, the engine was an integral part of the chassis). The belt pulley was mounted on the end of the crankshaft so power could flow directly from engine to belt-driven machine.

Thermo-syphon cooling eliminated the need for a water pump.

These engines burned low-cost fuel—distillates and various fuel oils.

All of these factors added up to one important feature—economy. During the Great Depression, this was most important.

Rugged engines? Yes. Crude? By no means! These engines were tough, but sophisticated as well. Over the years, they introduced turbulence-producing combustion chambers for a better mixture of fuel and air and better fuel economy. They were the first with a 2-position manifold; in cold weather, it could be set to "hot", diverting exhaust gases around the fuel-air passages to heat the mixture and assist combustion. Other features included positive lubrication, forced crankcase ventilation, an oil-bath air cleaner, natural-draft carburetion, and many others.

With all these advantages, why were these engines dropped?

Two-cylinder engines big enough to meet demands for increased power would require tractors with physical size too large for working between rows. These engines were not suited for the shift-up, throttle-back, save-fuel operation desirable for the lighter work which most tractors perform during more than half their working hours. In short, these great engines of the 20s to the 50s were simply not flexible enough for the 60s and 70s.

On the other hand... The cutaway drawing at left shows a 2-cylinder Model "D". Many "D"s are still at work on American farms, still earning their keep. Some of their owners will still give you an argument as to whether the 2-cylinder engine is obsolete.

"AO"—1936-1953

While the "A" swept the row-crop areas, other farmers asked for tractors which would do for them what the "A" was doing for row-crop farmers. Result: The first of several variations on the "A"—the "AR", with fixed (standard) tread, and the "AO", left, with stack removed from the hood and wheels shielded for work among trees. Both were introduced in 1935, for the 1936 model year.

Note the wheels on the "AO". John Deere first offered rubber tires on the "A" and its variations. Farmers were skeptical—afraid they'd lose traction with rubber—but extensive field demonstrations and the greatly improved comfort soon changed their minds. Tires then caught on rapidly.

Other "A" variations included the "AN", with narrow front end, and "AW", with wide front end, both introduced in 1936; and two high-crop models, "ANH" and "AWH", introduced in 1937.

The "AR" was rated at 20.35 drawbar and 26.30 belt hp. See State of Nebraska Test No. 378 (1941). Generally, variations on the "A" were rated the same as the "A" (see "A" ratings).

Just as wheatland farmers had admired the "A" but asked for a standard-tread version because they did not need adjustable wheels, small row-crop farmers admired features of the "A" but asked for a smaller tractor, more suited to their power needs. In 1935, John Deere answered with the "B".

Literature of the day described it as "... two-thirds the size of the "A" in power and weight. It has all of the advanced John Deere General Purpose features—adjustable wheel tread, ample clearance, excellent vision, easy steering, convenient controls, individually controlled rear wheel brakes, comfortable seat, roomy platform and four forward speeds."

Like the "A", the "B" had many variations. One was the "BN" (right), with a narrow front end.

The "B" was originally rated at 9.28 drawbar hp and 14.25 belt hp. See State of Nebraska Test No. 232 (1935); also No. 305 (1938), No. 366 (1941), No. 380 (1947), and No. 381 (1947).

"B"—1935-1952

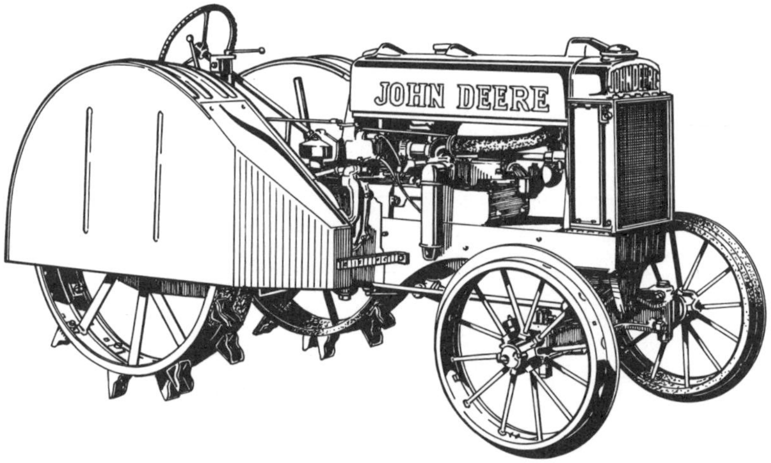

"BO"—1936-1947

Inevitably, introduction of a new row-crop tractor brought demands for variations. The wheatlands wanted a standard-tread model; orchardists wanted this plus shielding to protect trees. The "B" was no exception. A year after its introduction came the predictable "BR" and "BO", in 1936.

Note the fenders shown on the "BO" (left). These are citrus-type fenders to protect lush, low-hanging branches of citrus trees. Such fenders were also available for the "AO".

For power ratings on all variations of the "B" see the same State of Nebraska Tests listed for the "B".

"G"/"GM"/"G"—1938-1953

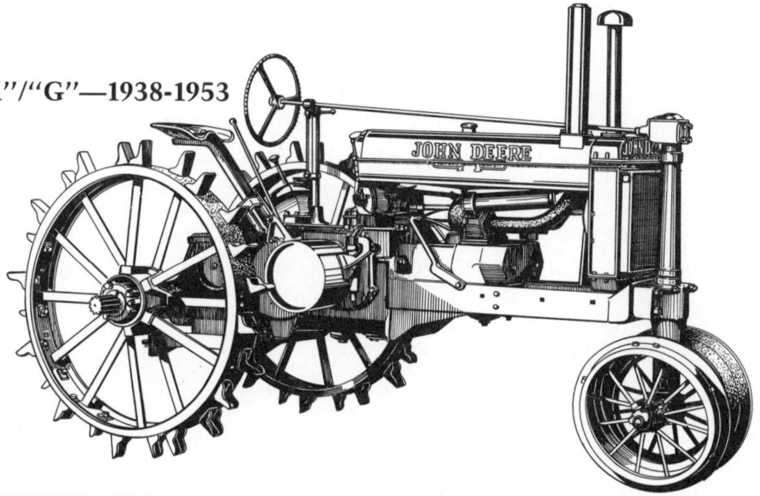

Power farming increased the working capacity of farmers and led to increasingly larger farms. Larger farms, in turn, required larger tractors. Noting this trend, John Deere introduced the Model "G"—the most powerful tractor it had manufactured until that time—in 1937.

The "G" could handle three 14-inch plows, a 10-foot single disk, 3-row bedders under the toughest conditions and 4-row bedders in most conditions, and 4-row planters and cultivators. Its belt could power a 28-inch thresher. A power take-off was standard equipment. Its power, combined with the other John Deere Row-Crop Tractor features, made it popular with large-acreage cotton and corn growers, and with grain farmers who also grew a few acres of the more common row crops.

The "G" was originally rated at 20.70 drawbar hp and 31.44 belt hp. See State of Nebraska Test No. 295 (1937); also No. 383 (1947).

"Utility tractor" describes a small tractor that performs a variety of "chores." It may be the secondary tractor used for work not requiring the horsepower available in larger, primary tractors on large farms. Or, it may be the primary tractor on smaller farms.

The Model "L", introduced in 1937, was the first John Deere Utility Tractor. It had a 2-cylinder engine but, unlike that in other John Deere Tractors of the time, it was a vertical, John Deere-Hercules engine. It had a foot clutch. An offset engine aided visibility while cultivating single rows. John Deere Wagon Works, Moline, built the "L".

In 1941 the "L", which could pull a one-bottom 12-inch plow in most conditions, was joined by the "LA", which pulled one 16-inch bottom.

The "L" was rated at 7.01 drawbar hp and 9.27 belt hp. See State of Nebraska Test No. 313 (1938). The "LA" was rated at 10.46 drawbar hp and 12.93 belt hp. See State of Nebraska Test No. 373 (1941).

"L"—1937-1946; "LA"—1941-1946

Early tractor design objectives emphasized power, versatility, dependability, economy. In the 30s, another was recognized: styling. It became axiomatic that well-engineered products also looked good. Appearance became a yardstick of quality.

John Deere retained a noted industrial designer, the late Henry Dreyfuss, to ensure the desired marriage of esthetics and performance. In 1938 came the first results of his efforts—new styling for the "A" and "B" (left).

This styling was adopted for the "D" in 1939 and the "G" in 1942. New tractors introduced after this styling wore the same "family look" until subsequent restylings.

Despite the increasing popularity of tractors, many farmers kept their horses until well through the 30s. Some plowed and planted with tractors but preferred Ol'Dobbin for the meticulous work of cultivating. Others, on small farms, believed the size of their operations simply did not justify a tractor. So the horse remained an important source of farm power. But in 1939, John Deere introduced the Model "H"... and the days of the draft horse were numbered.

The "H" was a small tractor and met the small farmer's needs for power at a price he could afford. It was welcomed on large farms as well, as an economical source of auxiliary power on lighter jobs that did not require all the power of the larger, primary tractor. Finally, for cultivation, it had a narrow, tapered hood for unimpaired visibility, and a complete line of matched working equipment including 2-row cultivators.

The "H" was built along the same general lines as the "A", "B", and "G", and shared many features with those larger row-crop tractors. Four variations of the "H" were produced: The original "H" (below); the "HN", with single front wheel; "HWH", wide front, high clearance; and the "HNH", single front wheel and high clearance.

The "H" was rated at 9.68 drawbar hp and 12.97 belt hp. See State of Nebraska Test No. 312 (1938).

"H"—1939-1947

"M"—1947-1952

During World War II, only variations on existing models were introduced—"LA" (1941); "GM", or styled "G" (1943); "A", "B", and "H" with narrow front end and high clearance (1944). 1947 saw significant improvements—the first fuel option, all-fuel or gasoline engines for the "A" and "B"; Powr-Trol (an industry first) for remote control of towed implements; electric starting and lighting systems as standard, built in at the factory; interchangeable front ends, including "knee action" Roll-O-Matic which made both steering and riding easier.

That same year saw the first tractor produced at a new factory in Dubuque, Iowa. It was the "M", (left) a general-purpose utility tractor. The "M" had many of the features mentioned already, plus one innovation all its own: Touch-O-Matic, which utilized tractor hydraulics to take the heavy work out of attaching and controlling integral equipment.

The "M" was rated at 14.39 drawbar hp and 18.21 belt hp. See State of Nebraska Test No. 387 (1947).

The first variation on the "M" was the "MT", or "M"-Tricycle, available with single front wheel, as shown, dual front wheels (double knuckle), "knee action" Roll-O-Matic, or adjustable front axle.

The "MT" also featured 2-lever implement control—Dual Touch-O-Matic—for independent adjustment of tools working on either side of the tractor. Pulling the right lever raised the right cultivator rigs, pulling the left lever raised the left rigs, pulling both simultaneously raised both right and left sides. Additionally, right and left arms could be pinned together and a single lever used to activate the power of both circuits to raise equipment. The levers could be used to control rear-mounted as well as front-mounted equipment.

The "MT" was rated at 14.08 drawbar hp and 18.33 belt hp. See State of Nebraska Test No. 423 (1949).

"MT"—1949-1952

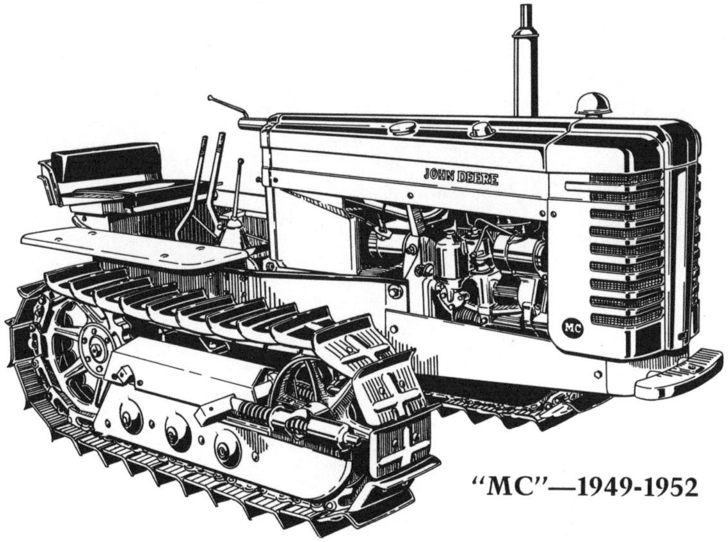

"MC"—1949-1952

Which was the first John Deere Crawler Tractor? You have a choice...

From 1940 until 1947, "BO" Tractor chassis were shipped to a factory in Yakima, Washington, to be mounted on crawler tracks, which orchardists of the Pacific Northwest favored in that day. The Yakima factory was owned by two brothers named Lindeman. Logically enough, those modified "BO" Tractors came to be known as John Deere Lindeman Crawlers.

The first totally John Deere crawler was the "MC", which was manufactured at the Dubuque factory. The "MC" had 12-inch tracks, standard; 10-or 14-inch, optional. Tread widths: 36 to 42 inches.

Those Northwestern orchardists liked the"MC" for its hillside stability. Crawler traction and flotation also meant it could work in mud and snow, and those capabilities made the "MC" a popular logging tractor in New England and swampy Southern timber regions.

Besides its use in the Far West, the "MC" became popular for general farming in areas where crawlers were preferred.

The "MC" was rated at 13.70 drawbar hp and 18.89 belt hp. See State of Nebraska Test No. 448 (1950).

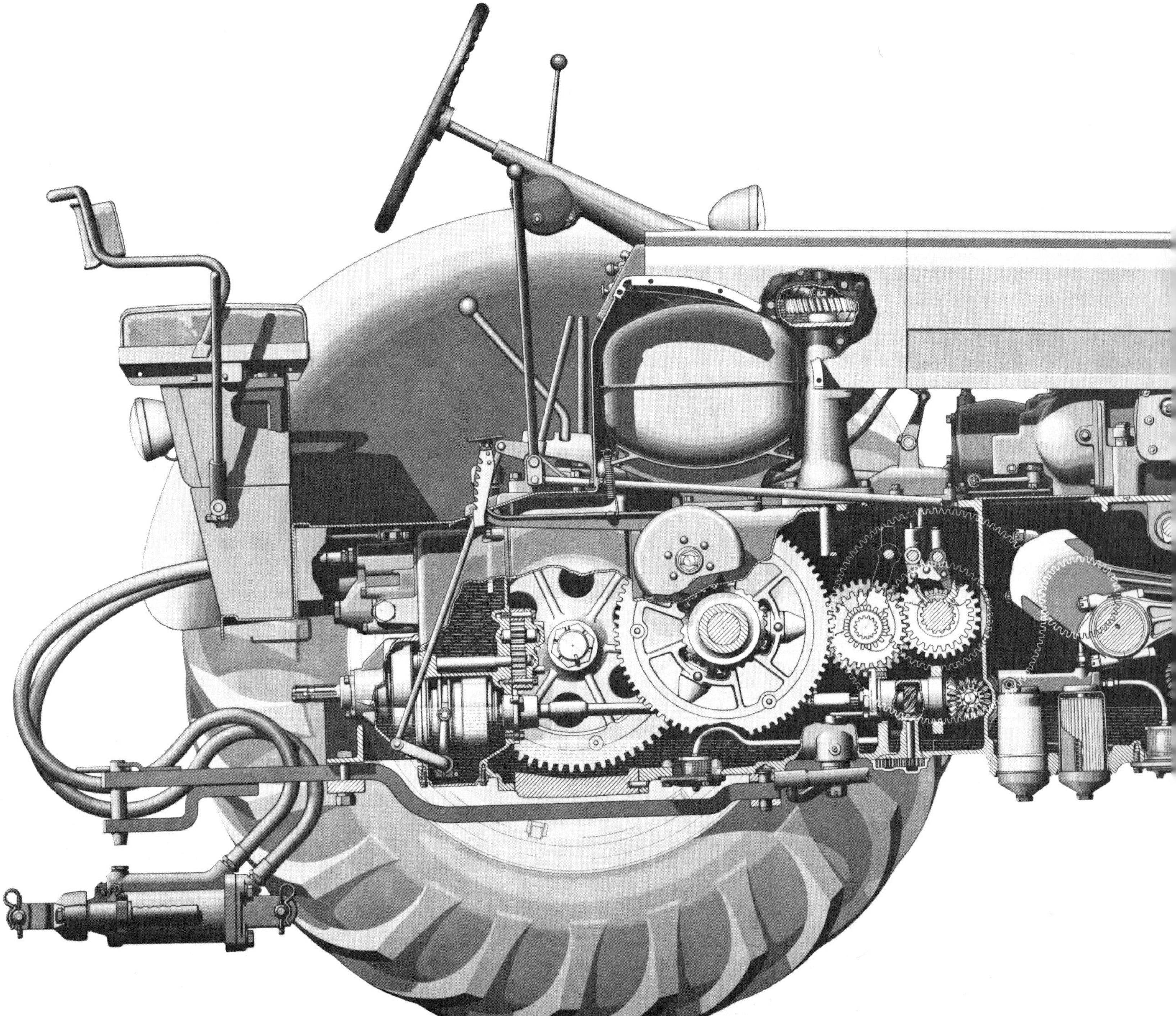

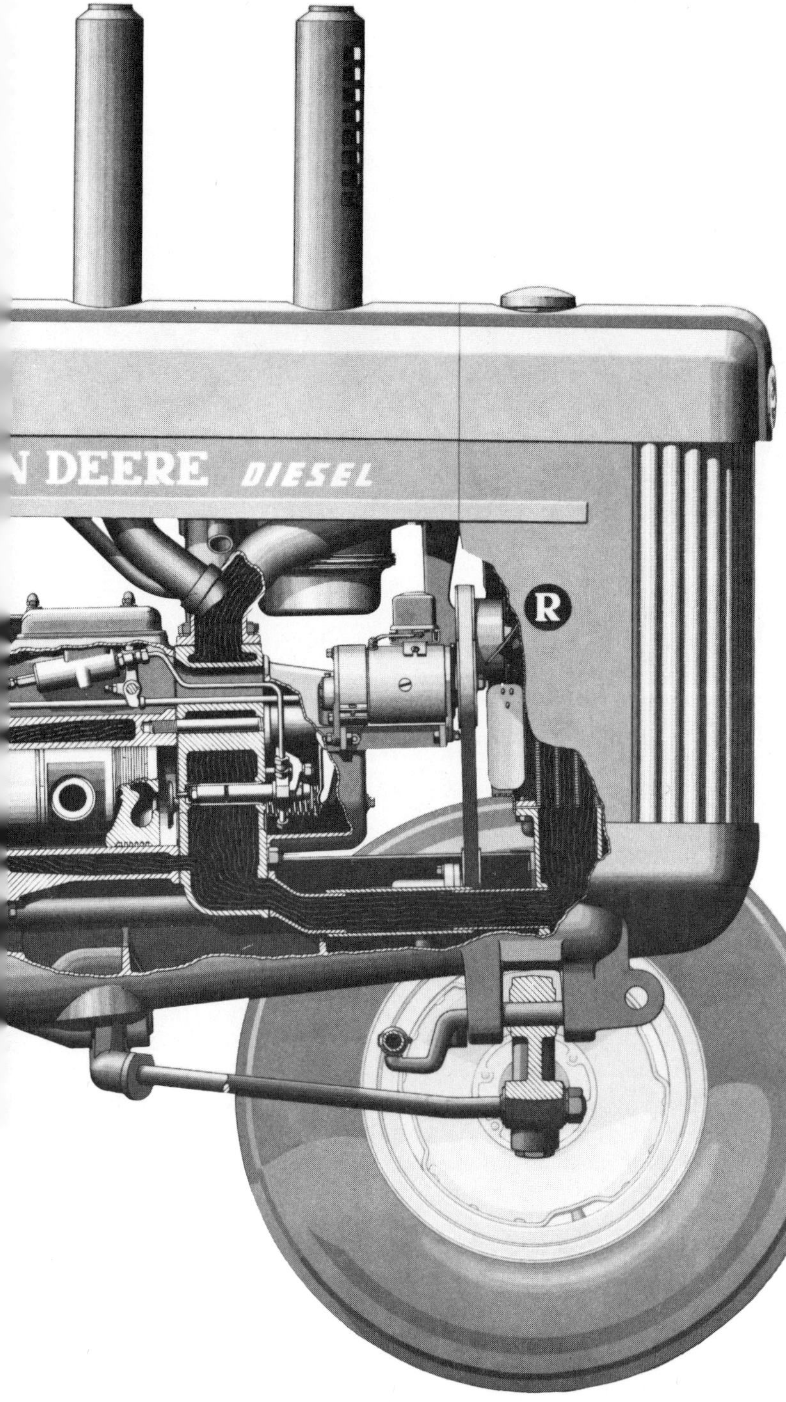

For well over two decades, while John Deere engineers developed a full line of row-crop tractors and their variations, the "D" was carrying the load in rice country and wheatland. By the mid-40s it was a much-changed tractor from the original "D"—mounted on rubber in the early 30s; styled in 1939; drawbar horsepower increased to 24.02 in 1935, 30.46 in 1940. But wheatland farms were growing larger, too, and demanding a standard tractor with even more power than the later models of the "D".

Coincidentally, a new fuel was becoming increasingly popular for industrial uses: Diesel. And diesel engine technology was reducing the disadvantages which had slowed the adoption of diesel power for agricultural tractors.

In the early years of World War II John Deere engineers began designing a standard tractor to replace the "D". Two of their goals were: (1) more power, and (2) a diesel engine. By 1949 they had realized both objectives in the Model "R" (left and below).

The "R" had a gasoline-fueled auxiliary starting engine, which eased two problems that had plagued earlier diesels. While others were started with the operator standing on the ground manually turning a flywheel, the "R" was started with the operator seated, pulling a lever. While others were slow to warm up, starting-engine heat enhanced warmup of the "R" for a fast start on the day's work, even in cold weather.

The "R" boasted many other features pioneered on John Deere Tractors, of course. Plus an all-steel cab, the first for a John Deere Tractor, available as extra equipment.

And the "R" had power—more power than any tractor John Deere had manufactured. It was rated at 34.27 drawbar hp and 43.32 belt hp. See State of Nebraska Test No. 406 (1949).

"R"—1949-1954

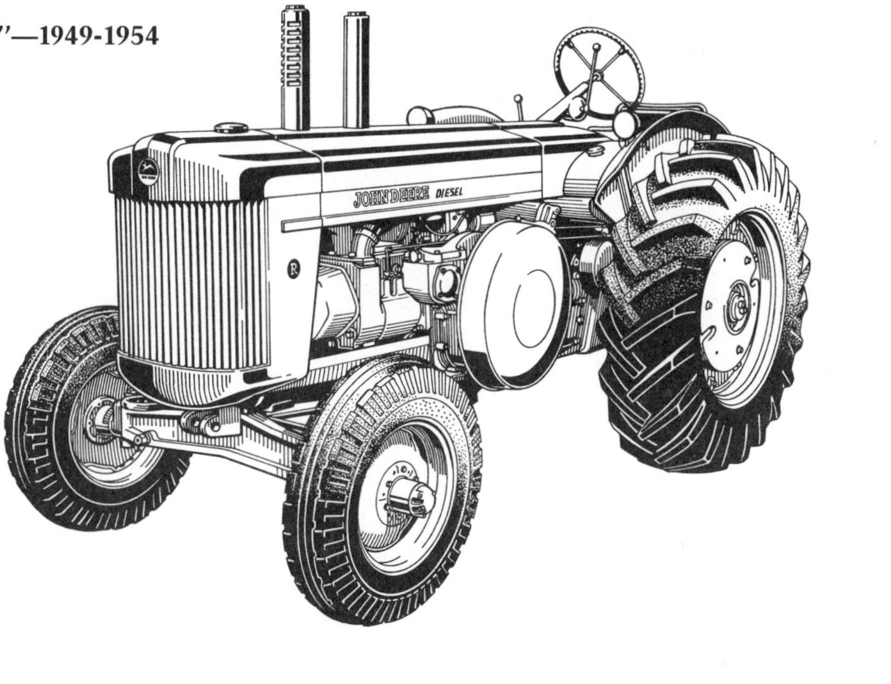

50—1952-1956

60—1952-1956

The first numbered tractors in the John Deere line, the 50 (above) and 60 (above right), arrived in 1952 as replacements for the "B" and "A" respectively. Both represented broad improvements in performance and convenience.

These were the first tractors with duplex carburetion (a separate carburetor for each cylinder), to meter fuel precisely for improved fuel efficiency and increased horsepower.

These were the first John Deere Tractors with a "live" powershaft, which provided continuous power for operating PTO-driven equipment whenever the tractor engine was running. (See more complete explanation on the following page.)

"Live" high-pressure Powr-Trol on the 50 and 60 further enhanced the already enviable reputation of John Deere hydraulic systems. It provided increased hydraulic power to handle heavier drawn or integral equipment and operated independently of both the transmission clutch and power takeoff.

Convenience features included an improved steering system, adjustable backrest on the seat, longer hand clutch and throttle levers to put these controls within easier reach of the operator. The air intake stack was under the hood, and the exhaust stack was offset on the hood to improve straight-ahead visibility. Changing rear wheel tread became a quick and easy operation with a significant industry first—rack-and-pinion rear wheel tread adjustment. Introduced on the 50 and 60, it remains on the most modern John Deere Tractors today.

The 50 was originally rated at 20.62 drawbar hp and 26.32 belt hp. See State of Nebraska Test No. 486 (1952); also No. 507 (1953), and No. 540 (1955).

The 50 and 60 were the first John Deere Tractors to offer farmers a choice of gasoline, "all fuel", or liquefied petroleum (LP) gas engine, the latter at a slightly higher cost.

LP gas was popular in a few localized areas. Near oil refineries it could be purchased for as little as 5 cents per gallon, making it economical even though a gallon of LP would not produce quite the work output of a gallon of gasoline. LP is a clean-burning fuel, reducing engine deposits. LP must be handled with pressurized equipment, but some farmers found this a significant advantage over gasoline, which was vulnerable to thieves with siphon tubes.

The 60 (and the 70, which joined the line a year later) offered one more option of particular value to producers of tall, bushy, bedded crops like sugarcane, flowers, etc. They were available as Hi-Crop tractors with at least 32 inches of clearance under both front and rear axles and a minimum of 48 inches between final drive housings. A special line of John Deere matched working equipment was available for use with these Hi-Crop models.

The 50 and 60 also offered a variety of front-end and rear-axle options to fit virtually any row-crop need: Single front wheel, Roll-O-Matic, 38-inch fixed-tread front axle, or front axle adjustable in four steps from 56 to 80 inches. A long rear axle option permitted 62- to 97-inch rear wheel tread.

The 60 was originally rated at 27.71 drawbar hp and 35.33 belt hp. See State of Nebraska Test No. 472 (1952); also No. 490 and No. 513 (1953).

70—1953-1956

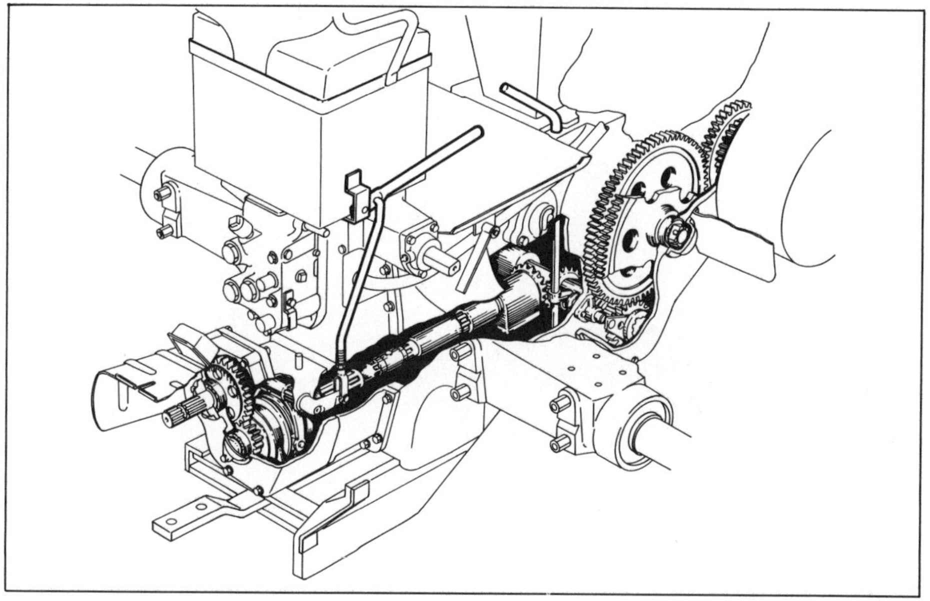

The 70 arrived a year after the 50 and 60, with many of the same features. Later it earned two distinctions all its own, and shared others with its year-older brothers.

Originally available with gasoline, "all fuel", or LP-gas engine, it later offered a diesel option. The 70 thus became the first John Deere diesel row-crop tractor.

In its Nebraska test, the 70 Diesel set a new fuel economy record, bettering all previously tested row-crop tractors. In 1954, the 50, 60, and 70 became the first row-crop tractors equipped with optional factory-installed power steering. John Deere engineers achieved this industry first with a system that used built-in hydraulics to control the steering column. This differed from "add-on" systems utilizing externally mounted motors on steering shafts or hydraulic cylinders hooked up to tie rods.

Improvements in equipment control paralleled improvements in tractor power and performance. Case in point: The 801 Hitch, an early weight-transfer hitch that transformed implement draft resistance into downward pressure on tractor drive wheels. It was introduced on 50, 60, and 70 Tractors.

In addition to the row-crop model, the 70 was available as a somewhat unusual standard tractor. More accurately, it was a wheatland tractor built on a row-crop chassis. Fenders and front axle were all that distinguished it from the row-crop model.

The 70 Diesel was rated at 34.25 drawbar hp and 43.77 belt hp. See State of Nebraska Test No. 528 (1954); also No. 493, No. 506, and No. 514 (1953).

Tractor power is used for many purposes other than pulling. The early tractors had a pulley, mounted on an extension of the engine crankshaft, so belt power could be "taken off" the engine to power threshers and other machinery. Later, a system was developed to transmit power through a shaft protruding through the rear of the tractor. It was quickly named a power takeoff, or PTO.

Early PTO use was limited. Transmission-driven, the PTO could be used conveniently for such stationary operations as powering hammermills or stationary hay balers. But field operation with corn pickers or forage harvesters was awkward and difficult. The continuous-running PTO, controlled by a dual-stage clutch pedal that also controlled the transmission, was an improvement. But maximum utilization of the PTO awaited the fully independent PTO controlled by its own clutch.

Such a PTO was introduced to John Deere Tractors on the 50, 60, and 70. The operator could start or stop tractor travel, shift gears, or operate the hydraulic system without affecting PTO operation. The independent PTO (pictured in the cutaway drawing above) popularized PTO work and greatly extended the practical uses of tractor power.

40—1953-1955

40C—1953-1955

Introduced in 1953, the 40 Standard, 40 Tricycle (above), and 40C Crawler (right) replaced the "M", "MT", and "MC".

The 40 Series Tractors were Dubuque-built, smaller brothers of the 50, 60, and 70. Family resemblance was most evident in the styling, but, as the advertising of the day said, "the resemblance is more than skin deep." Features similar to the larger, Waterloo tractors included an improved 3-point hitch; "live" hydraulic power for easy raising, lowering; and precision setting of working equipment; and Touch-O-Matic, which allowed individual control of cultivator sections right and left.

Popular features retained from the "M" Series included four forward speeds; adjustable wheel tread; power takeoff; and self-energizing, independent rear-wheel brakes for sure stops and short turns.

Compared with the discontinued "M" tractors, both the 40 Standard and 40 Tricycle tractors had larger fuel tanks and a new radiator grille with a deep-fluted, easy-to-clean screen ahead of the radiator. Servicing of the battery was made easier with a more practical design of the battery compartment. These tractors were also easier to mount and dismount and had an adjustable seat with coil-spring cushion.

40 Series tractors also boasted approximately 15 percent more power than their "M" Series predecessors.

The 40 Tricycle model was rated at 17.16 drawbar hp and 21.45 belt hp. See State of Nebraska Test No. 503 (1953); also No. 504 (1953) and No. 546 (1955).

The 40C Crawler, introduced in late 1953, offered the same increase in power, operator comfort, and styling advantages as other 40 Series Tractors. And it had one other significant advantage... an ever-widening market.

The Pacific Northwest orchardists who pioneered the use of small crawlers there did so because crawlers provided hillside stability. And the good flotation appealed to New England maple sugar producers who tended their trees while snow still covered the ground; Southern foresters who cut timber in swampy woodlands; builders who needed a tractor to scoot over mounds of freshly excavated soil; and those farmers who needed to work in fields too soggy to support wheeled tractors.

The 40C had a desirable balance of horsepower, weight, and ground speeds. This balance, and the approximately 15 percent power boost over the "MC", made it suitable for many applications beyond the capacity of the "MC". A tool carrier specially designed for John Deere Crawlers permitted easy attachment of plows.

Note the radiator shield on the 40C pictured above. This was considered a necessary option in woodlot, orchard, and forest operations to protect the radiator and lights from branches.

The 40C was rated at 15.11 drawbar hp and 21.24 belt hp. See State of Nebraska Test No. 505 (1953).

Six years after the introduction of the historic "R", its replacement arrived in the 80 Diesel. It was big, weighing 7,850 pounds (shipping weight) stripped down, and 8,150 pounds when equipped with options. It was powerful... 57.49 hp at the belt pulley, 46.32 hp at the drawbar.

It incorporated many of the most popular features introduced on the 50, 60, and 70 Tractors, such as optional power steering, "live" powershaft, Powr-Trol, and a 6-speed transmission.

The 80 Diesel was capable of pulling a 21-foot disk and working up to 125 acres with it in a single day. A big 32½-gallon fuel tank allowed the operator to spend a long time in the field between refueling stops. It had an electric pushbutton fuel gauge.

Technically, the 80 Diesel was a standard tractor with a rear wheel tread of 64½ inches (14-34 or 15-34 tires) or 67 inches (18-26 tires). But these tread widths could be increased to 68½ and 75 inches simply by reversing the rims of the wheels. Up front, an oscillating axle cushioned shocks and provided flexibility for working rough ground. The front tread could be changed from 56½ inches to 63 inches by reversing the rims.

The 80 Diesel was a compactly designed tractor with an 85-inch wheelbase. This provided a high degree of maneuverability.

It was efficient, too. When tested at Nebraska, it set what was then an all-time record for diesel fuel economy.

The 80 was rated at 46.32 drawbar hp and 57.49 belt hp. See State of Nebraska Test No. 567 (1955).

80 Diesel—1955-1956

1956 saw the replacement of the 50, 60, 70 family of tractors with a new family, the "20" Series. These tractors incorporated advancements in virtually every aspect of tractor design and engineering.

Hydraulics—Custom Powr-Trol was introduced on the "20" Series. Raising and lowering of equipment was accomplished through a position-responsive rockshaft, which enabled the operator to set working depth precisely. An adjustable stop let him return implements to a preset working depth after raising them. Exclusive Load-and-Depth Control kept implements working at a uniform depth, automatically transferring weight to tractor drive wheels as needed to maintain traction. In extremely tough conditions, the system raised the implement just enough to relieve the load and allow the tractor to move steadily.

Engines—All new. Improved cylinder head and pistons increased combustion-chamber turbulence and increased horsepower. These engines set five new fuel economy records at Nebraska.

Operator comfort and convenience—The "20"s introduced a new Float-Ride seat, supported on a rubber torsion spring, with tension adjustable to match each operator's weight. A hydraulic shock absorber mounted between seat and seat pedestal also helped smooth the ride over rough ground. Foam-rubber seat cushion and backrest, and optional cushioned armrests, added still more comfort. There was a roomy stand-at-will platform, easy-to-reach controls, and easy-to-read instrument panel as well.

Styling—The first John Deere Tractors to wear a two-tone paint job.

The 520 was rated at 25.63 drawbar hp and 32.38 belt hp. See State of Nebraska Test No. 590 (1956); also No. 592 and No. 597 (1956).

The 620 was rated at 34.34 drawbar hp and 42.79 belt hp. See State of Nebraska Test No. 591 (1956); also No. 598 and No. 604 (1956).

The 720 was rated at 40.63 drawbar hp and 50.67 belt hp. See State of Nebraska Test No. 593 (1956); also No. 594, No. 605, and No. 606 (1956).

520—1956-1958

720—1956-1958

620—1956-1958

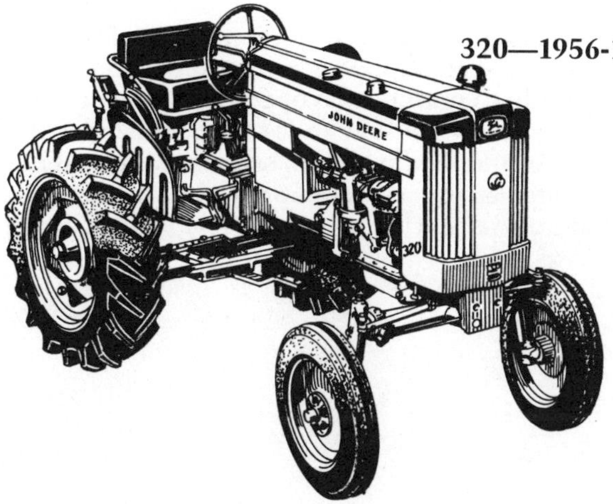

320—1956-1958

It was predictable that farmers who admired features in a large tractor would ask for these same features in smaller tractors more closely matched to their power requirements. John Deere anticipated these demands in the mid-50s, and planned the total "20" Series package to include two utility models, the 320 (left) and 420 (below).

Like the bigger "20" Series Tractors from Waterloo, the 320 from Dubuque had the new 3-point hitch with Load-and-Depth Control, Touch-O-Matic hydraulics, adjustable coil-spring cushion seat, and independent disk-type brakes. Starting was easy—with a pushbutton.

Special equipment for the 320 included a lighting package with two forward lights, one rear, and an instrument panel light; auxiliary hydraulic outlet for one remote cylinder; a rear wheel jack; and a PTO shaft.

The 320 was available as a standard or utility tractor. It was rated as a 1-or 2-plow tractor—unofficially, since it was never tested at the Nebraska Test Station.

The 420 Tractor came set up just about any way a buyer wanted. In fact, there were so many 420 conformations that the advertising literature of the day referred to the "420 fleet" rather than a single model.

There was a 420 General-Purpose Tractor with dual-wheel tricycle front-end, another essentially the same with single-wheel front end, a third with wide-axle front end. There was a 420 Standard. There was a 420 Hi-Crop with 32 inches of under-axle clearance, and a 420 Special with 27-inch clearance over the row and a wide range of wheel spacings. There were two 420 Utility models, one which straddled and cultivated two rows, another with a low profile for grove, orchard, and vineyard work. Finally, there were two 420 Crawler options, one with a 4-roller track system, the other with five rollers to provide maximum flotation, traction, and fore-and-aft stability.

The 420 was rated at 20.31 drawbar hp and 24.83 belt hp. See State of Nebraska Test No. 599 (1956); also No. 600 and No. 601 (1956).

420—1956-1958

820 Diesel—1956-1958

The 820 Diesel was a continuation of the diesel-powered, standard tractor tradition established by the "R" and continued through the 80. In fact, it was originally powered by the same engine as used in the 80. Later, its power was increased and it was sent to Nebraska for a new test. (The Nebraska Test information cited below refers to that later test.)

The 820 handled a 21-foot disk with ease. Likewise, it had plenty of power to pull multiple-drill hookups or field cultivators as wide as 24 feet.

A 6-speed transmission allowed the 820 to operate at speeds from 2⅓ to 12½ mph. Additionally, an optional "creeper" first gear could be installed to provide a 1¾ mph speed, ideal for maintaining full powershaft output when operating power-driven equipment in heavy crops, one example being its use with early power-driven, pull-type combines.

The 820 Diesel was rated at 52.25 drawbar and 64.26 belt hp. See State of Nebraska Test No. 632 (1957).

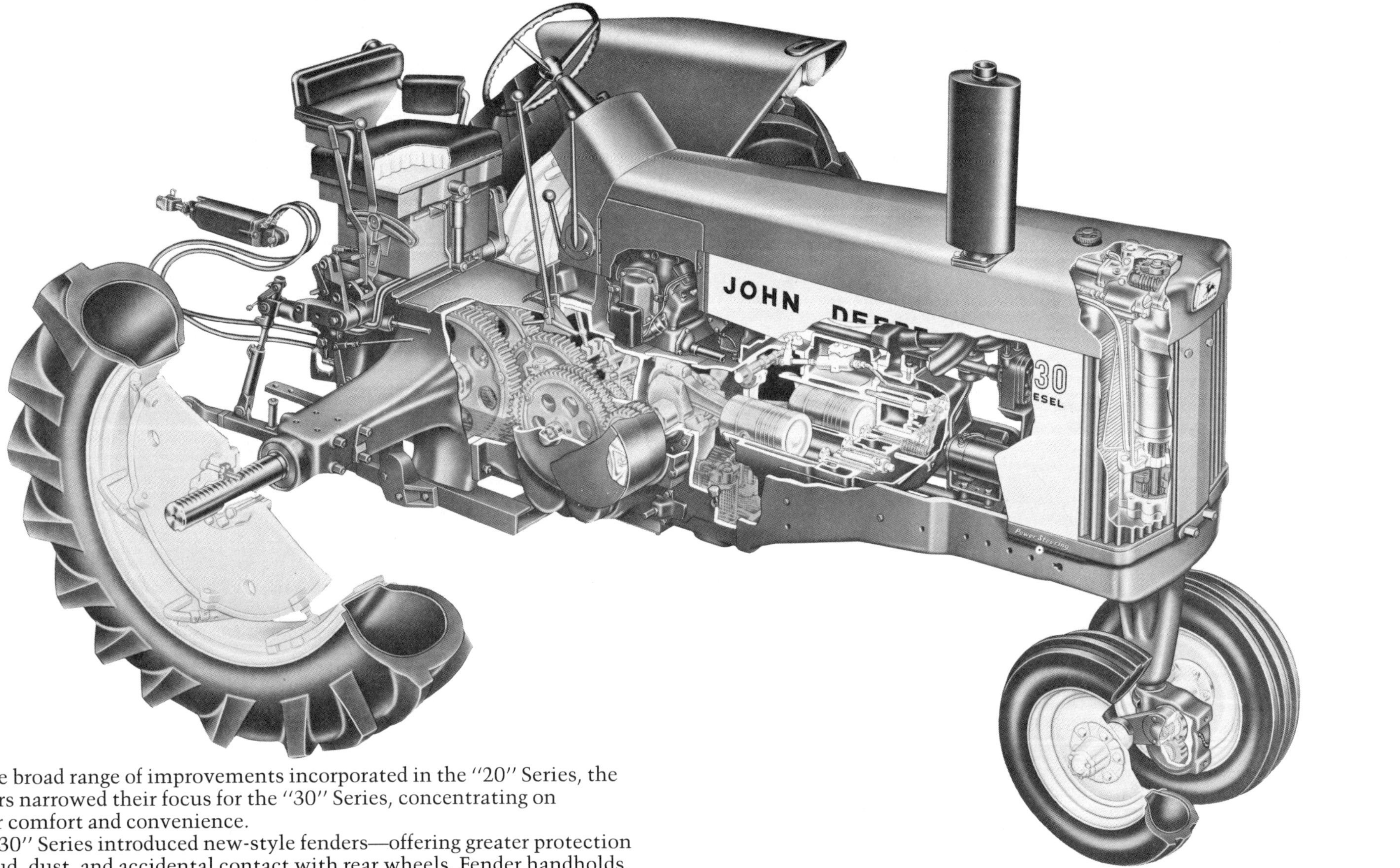

After the broad range of improvements incorporated in the "20" Series, the engineers narrowed their focus for the "30" Series, concentrating on operator comfort and convenience.

The "30" Series introduced new-style fenders—offering greater protection from mud, dust, and accidental contact with rear wheels. Fender handholds, and a step mounted on the rear axle, made mounting and dismounting easier. Once aboard, the operator sat comfortably in a deep-cushioned, adjustable seat with cushioned backrest.

Dual, sealed-beam lights were mounted in each fender. Inner lights projected ahead, outer lights flooded the working area.

Steering was redesigned, with the wheel mounted on a column projecting upward from the instrument panel, permitting a tilted wheel more natural to grasp.

A new instrument panel put fuel, oil-pressure, ammeter, and coolant temperature gauges in an easy-to-read cluster around the steering column. Speed-hourmeter, cigaret lighter, ignition switch, and starter button were in a lower panel. Both panels were illuminated for night work.

Starting became a pushbutton matter, the button located conveniently just below the ignition switch. The 730 Diesel introduced electric starting for diesels, a simpler and lower-cost alternative to the diesel starting systems available earlier.

None of the two-cylinder "30s" were tested at Nebraska. Since these tractors had the same engines as their "20" Series predecessors, a certification from the Test Station permitted "20" Series test results to be used for their "30" Series counterparts.

The 330 (right) and 430 (below) were the utility tractors in the "30" Series, but were also available in other configurations: The 330 as a standard tractor for one- or 2-row farming, the 430 as a row-crop. These Dubuque-built tractors incorporated many of the convenience features from the bigger Waterloo "30" Series, including the slanted, easy-to-read instrument panel; tilted steering wheel; adjustable, deep-cushioned seat. Familiar names and terminology were found among the list of options—Touch-O-Matic; 3-point hitch with Load-and-Depth Control; power-adjusted rear wheels.

The 330 and 430 were powered by John Deere-built, 2-cylinder, valve-in-head engines. The 330 was available as a gasoline model only. It had a 4-speed transmission, with speeds from 1⅝ to 12 mph.

The 330 had ample power to pull two 12-inch bottoms in most soil conditions and "approximately 21.5 rated belt horsepower."

330—1958-1960

430—1958-1960

The 430 had all the features of the 330 and a good deal more. One important "plus" was a time-saving, clutch-type direction reverser. This factory-installed option permitted use of any of the forward-gear speeds for reverse without shifting transmission gears. It was especially handy for that time-honored utility tractor chore, cleaning manure from barns and feedlots. A front-end loader could be lowered to the ground as the tractor moved forward to pick up a load, then operating the direction reverser was all it took to back away, move forward to dump into the spreader, then shuttle back to start the whole cycle again.

The various configurations of the 430 included a 430 Crawler, available as a 4-roller or 5-roller model.

The 430 had many uses: It could pull a 3-bottom plow or an 8-foot disk in most conditions. It could be used for seedbed preparation, haymaking, spraying, and pulling small harvesters. The gasoline version was advertised at 27.08 drawbar hp and 29.21 belt hp.

The 530 (right) was the smallest of the Waterloo introductions of 1958, but it had all of the deluxe features of the bigger tractors. Thus, it was a happy marriage of big tractor features and small tractor economy.

The 530 had plenty of power to handle three plow bottoms in most common field conditions, or any of a variety of drawn, integral, power-driven, or 3-point-hitch equipment. It satisfied all the power requirements on the medium-size row-crop farms of that day and could handle a goodly share of the work on larger operations.

The 530 was available with gasoline, LP-gas, or all-fuel engine; with conventional or Roll-O-Matic dual front wheels, single front wheel, or a 48- to 80-inch adjustable front axle. These front-end assemblies were easily interchangeable.

Advertised horsepower for the gasoline version: 34.31 drawbar hp and 38.58 belt hp.

530—1958-1960

630—1958-1960

The 630 (left) was described as a "heavy-duty 4-plow tractor that meets the power requirements of medium- and large-acreage farming operations." It had enough horsepower to handle a fully integral 4-bottom plow easily in most conditions or to work the then-new John Deere 6-row planters and cultivators. It was capable of pulling heavy equipment through tough conditions and doing so in higher gears for increased overall output.

Typical work loads for the 630 in various gear ranges: 1st gear, slow or "creeper" work and heavy PTO work; 2nd, normal PTO work or heavy tillage work in difficult conditions; 3rd, heavy tillage work in normal conditions; 4th, lighter duty loads; 5th, light work such as rotary hoeing or transport of heavy equipment; 6th, transport of the tractor bare or with lightweight equipment.

Advertised horsepower for the gasoline version: 44.16 at the drawbar, 48.68 hp belt.

730—1958-1961

"One man becomes a giant in terms of daily work output with a new John Deere 730 Tractor." So said the literature of 1958, and the 730 proved to have the capabilities to make the slogan ring true. It was a big tractor—6,790 pounds with gasoline or diesel engine, 7,100 pounds with an LP-gas engine. It "breezed" through the fields with a 5-bottom plow, 18-foot disk harrow, 6-row planting and cultivating equipment, 5- and 6-row bedders.

This power made the 730 the most popular of the "30" Series Tractors. It was an era when farms were getting bigger, and the demand was for more power to keep pace with increased work loads.

Advertised horsepower for the gasoline version: 53.05 hp drawbar and 59.12 hp belt.

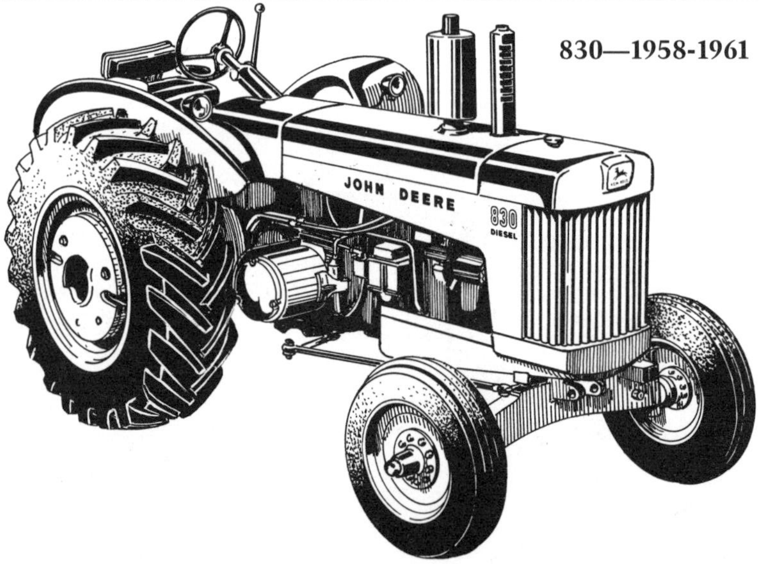

830—1958-1961

In its day, the 830 Diesel represented the ultimate in standard tractor power. Equipped with all the comfort and convenience features of the other "30" Series tractors, the 830 had the power to pull 5- and 6-bottom plows, 16-foot tool carriers, 20-foot disk tillers, 21-foot double-action disk harrows, multiple hookups of grain drills, field cultivators, and similar loads.

It was a big tractor, bigger even than the 730, with a shipping weight of 8,150 pounds—more than four tons—when equipped with power takeoff, Powr-Trol, and power steering.

The 830 offered its buyers a choice of starting systems—either the familiar V-4 gasoline starting engine or a convenient all-electric starting system powered by the tractor's 24-volt battery. Additionally, heating elements were available to assist cold weather starts when the 830 was equipped with the all-electric starting system. It also had a handy foot-operated accelerator-decelerator available as an accessory.

The 830 was advertised at 69.66 drawbar hp and 75.60 belt hp.

435—1959-1960

PTO use increased once it became independent (see page 17). But by the late 50s, new industry-wide standards were needed to handle increasing power and to make PTO-driven equipment interchangeable on more tractors. These came in 1958.

Lack of standardization in attaching points and PTO shafts had made some equipment incompatible with some tractors. In 1958, the farm equipment industry adopted a new standard that "...establishes the specifications that are essential in order that any 540-rpm PTO-driven machine may be operated with any make of tractor having a 540 rpm PTO drive." Specifics covered size and location of drawbar hitching holes, distance relationships between hitching points and PTO shafts, and such related matters as strength of drawbars, shaft dimensions and safety shielding.

That same year, a 1000-rpm PTO standard was adopted. Tractor horsepower and torque were increasing beyond that which could be effectively transmitted through the 540-rpm PTO established as an industry standard in 1923. At the higher speed, power-transmitting capacity was increased without increasing size of the PTO shaft.

The first John Deere Tractor introduced after the adoption of the new standards was the 435 Diesel Row-Crop Utility. It could be purchased with either 540- or 1000-rpm PTO. The Nebraska Test Station also had been watching PTO trends. For its 1959 tests, it dropped the belt-pulley rating and started PTO ratings. Thus, the 435 became the first John Deere Tractor to have its power reported officially in terms of drawbar and PTO power. The 435 was rated at 29.41 drawbar and 32.91 PTO hp. See State of Nebraska Test No. 716 (1959).

8010/8020—1960-1964

The first John Deere 4-Wheel-Drive Tractor was the over-200-hp 8010, introduced in the autumn of 1959. The occasion was a John Deere field day near Marshalltown, Iowa. Farmers were invited to examine new equipment which would be offered for the following crop year and witness its demonstration in the field.

To say the 8010 was the star of that show is an understatement. Onlookers were awed by this giant. It stood 8 feet 2½ inches tall to the top of its steering wheel, 8 feet wide, and 19 feet 7¼ inches from grille to drawbar. It weighed 19,700 pounds (24,860 pounds with its 23.1-26 12-ply tires ¾ full of ballast).

Four steps and a handhold were built onto each front fender to assist the operator in mounting to his spacious platform. A key and pushbutton, plus 24-volt electric system and fluid aid, provided fast starting in any weather. Articulated power steering and air brakes, both provided as regular equipment, kept the 8010 under easy control.

Performance? A 31-foot harrow and an 8-bottom plow were designed for the 8010. It pulled either at speeds up to 7 mph, and lifted the fully integral plow high off the ground for turns at row ends.

A year after the 8010 came the 8020, with an 8-speed Syncro-Range transmission, an improved posture seat, oil-cooled clutch, and other advances.

The 8010 and 8020 were never tested at Nebraska. Unofficially, they were reported at 215 engine hp and 150 hp at the drawbar.

New Generation of Power

Times change. Circumstances change. Problems change. And as they do, tried and true solutions must also change.

The 2-cylinder engine was "tried" over the course of nearly half a century, and for the reasons described on page 9 was found to be "true". But changing times had brought changing farms, bigger farms producing more crops, and requiring more power to do it. John Deere engineers had made 2-cylinder engines more powerful, but the power limits were being reached unless tractors were to grow so large they would be impractical for farming.

Fuels were also changing. The low cost of distillate fuels had been a major factor in the growing popularity of John Deere Tractors. But World War II innovations in petroleum refining ended that; distillates became as expensive as gasoline. In the early postwar years, gasoline became the preferred fuel, diesel was coming on strong, and distillates lost favor.

Recognizing these trends of the late 40s and early 50s, Deere & Company asked its tractor engineers to develop long-range plans. The first step was extensive field surveys to learn what farmers wanted. Results: Demands for more changes, revolutionary changes. Farmers asked for more power in nimble, responsive tractors that would let people farm faster. Transmissions with at least six speeds. Ability to put more horsepower through the powershaft. Bigger fuel tanks. More operator comfort. Easy-to-operate controls. Hydraulics that would do more. And all of this in tractors priced within reach of the farmers to be served.

Start from scratch? That seemed the only possible approach.

For a few years, Waterloo engineers pursued two objectives: Improve the current line and develop a new line. But demands of the first distracted from the second. So, in 1953, a team of engineers—engine, chassis, transmission, and hydraulics experts—moved from factory engineering offices to a leased building some distance away. Benefits: Isolation from day-to-day projects involving the current line, and secrecy for their work on the new line.

In 1960, the results of their efforts were introduced: Four new models (pictured left), described on the following pages.

In manufacturing circles, a model change that requires 33 percent new parts is considered a major change. This new line of tractors was made of 95 percent newly designed parts.

One of the tractors, the 4010 Diesel, would become the most copied tractor in the industry's history. Collectively, this "New Generation of Power" would virtually revolutionize farm tractor design. In fact, it would virtually revolutionize farming.

1010—1961-1965

Smallest of the "New Generation of Power" Tractors were the four models in the 1010 Series, described as "compact" tractors providing "A man's work at a boy's pay." The series included:

The 1010 Utility model—a handy, extra-low tractor, at home in vineyards, groves, and orchards as well as handling tillage, haying, hauling, loading and other chores, particularly on smaller farms.

The 1010 Row-Crop Utility—a New Generation successor to earlier tractors combining a low profile and a roomy under-axle clearance. It could cultivate standing crops and also worked in cramped quarters inside farm buildings or pens.

The 1010 Single Row-Crop— for speciality crop growers who worked one row at a time, particularly in tobacco and vegetables. Offset front wheels and a seat that could be offset provided a clear view of the row being worked.

The 1010 Crawler—for dozing, logging, excavating, land clearing, and subsoiling, in addition to vineyard and orchard work.

All 1010 models had variable-speed, 4-cylinder gasoline or diesel engines with 5-speed sliding-gear transmission on the wheeled models, and 4-speed sliding-gear transmission on the crawler. Row-Crop Utility and Single Row-Crop models were available with two independently controlled rear rockshafts. These models also had a 3-point hitch with Load-and-Depth Control sensing which could handle Category 1 equipment. The crawler had a dual-cylinder-operated 3-point hitch for Category 1 or Category 2 equipment.

The 1010 (Row-Crop Utility, diesel) was rated at 35.99 PTO hp and 30.82 drawbar hp. See State of Nebraska Test No. 803 (1961); also No. 798 (1010 C, diesel, 1961), No. 801 (1010 C, gasoline, 1961), and No. 802 (1010 Row-Crop Utility, gasoline, 1961).

2010—1961-1965

The 2010 could be ordered as a full-fledged row-crop tractor (above), as a multitalented row-crop utility, or as a high-clearance "Hi-Crop" tractor for working in tall or bushy crops. Power? Plenty for a 3-bottom plow or 2- and 4-row cultivators.

The Row-Crop Utility lived up to its name, with $19\frac{1}{2}$-inch crop clearance while standing only $55\frac{5}{8}$ inches tall at the hoodline. It was available with a swept-back front axle that reduced its wheelbase and gave it a shorter turning radius.

The 2010, along with the 3010 and 4010 (shown right), offered Syncro-Range transmission, which provided eight forward and three reverse speeds. Combined with the 2010's variable-speed 4-cylinder gasoline, diesel or LP-gas engine, Syncro-Range offered infinitely variable ground speeds up to about 20 mph in overlapping ranges. The operator used only the horsepower required for the job at hand; he could throttle down in higher ranges for greater economy on light work or throttle up and shift down for heavier jobs. Synchronizers in the transmission brought gears up to matching speed before engagement, permitting shifting on the go within a range between forward speeds or shuttle shifting between forward and reverse.

Independent 1000- or 540-rpm PTO was available.

The 2010 RU Diesel was rated at 46.67 PTO hp and 39.95 drawbar hp. See State of Nebraska Test No. 799 (1961); also No. 800 (2010 RU gasoline, 1961), No. 829 (2010 C gasoline, 1962), and No. 830 (2010 C diesel, 1962).

3010—1961-1963

The 3010 was equally at ease with 4-bottom plows, 4-row cultivators, 5-row planters, and similar big-capacity tools for large row-crop farms. The 3010 was offered as a row-crop, standard (above), or row-crop utility tractor.

In common with the 4010, the 3010 introduced closed-center hydraulics, one of the most historic innovations of the "New Generation of Power". The hydraulics experts on that engineering team installed a variable-displacement pump connected to a single reservoir, providing a hydraulic system which delivered instant power on demand for a variety of uses. The system provided up to three independent "live" hydraulic circuits to serve a rear rockshaft and one or two remote cylinders in addition to power steering and, on the 3010 and 4010, hydraulic power brakes, another industry first. One or two remote cylinders could be used to operate a front rockshaft to raise and lower front-mounted equipment and to provide down-pressure on that equipment if needed.

An exclusive engine balancer—two sets of gears, driven by the crankshaft and rotating in opposite directions at twice the speed of the engine—eliminated much of the vibration associated with conventional 4-cylinder engines.

The 3010 Diesel was rated at 54.36 drawbar and 59.44 PTO hp. See State of Nebraska Tests No. 762 (1960), No. 763 (gasoline, 1960), and No. 764 (LP-gas, 1960).

4010—1961-1963

The 4010 and other "New Generation of Power" tractors introduced the high-horsepower-to-weight concept to the John Deere line. Comparing the 4010 Diesel with the 730 Diesel illustrates this concept. The 730 had 53.66 drawbar hp; its shipping weight was 6,790 pounds. The 4010 had 73.65 drawbar hp, yet its shipping weight was only 6,980 pounds...37% more power, but only 3% more weight.

High-horsepower-to-weight tractors have many advantages. First and most obvious, a smaller percentage of the available power is required just to move the tractor. Second, this design encourages operation with equipment loads that can be pulled at higher speeds. The result is more productive use of power, reduced lugging strain on the engine, and increased efficiency. The combination of 6-cylinder variable speed 4010 engine and Syncro-Range transmission meant the operator could select the most efficient gear and speed for the job at hand.

And the 4010 offered many other desirable features, most in common with the other "New Generation of Power" Tractors. Performance: A 540/1000-rpm PTO, quickly convertible; new 3-point hitch with lower-link sensing; hydrostatic power steering; front rockshaft; and a new scientifically designed posture seat, the result of a cooperative effort between John Deere engineers and a physician who specialized in problems of the back. It was so comfortable that farmers sent fan mail praising it.

The 3010 and 4010 also introduced power brakes to farm tractors. The 4010 was available as a standard, row-crop or hi-crop, with gasoline, diesel, or LP-gas engine. The 4010 was originally rated at 84 PTO hp. See State of Nebraska Tests No. 761 (1960), No. 759 (1960), and No. 760 (1960).

5010—1963-1965

Two years after the introduction of the "New Generation of Power" came the first addition to that line. A big addition it was, the 5010; first 2-wheel-drive tractor with over 100 hp at both drawbar and PTO.

Big capacity? Yes. But a versatile and maneuverable tractor as well. And therein lies its reason for being. In the major small-grain producing areas, there was an obvious need for an extraordinarily powerful standard tractor to speed up work measured in sections rather than acres. And in hillside wheat areas, there was a long-felt need for a rubber-tired, faster-working, less-costly replacement for the lumbering crawler. The 5010 was it.

In addition to its power and speed, the 5010 had a new Quik-Coupler 3-point hitch to accommodate both Category 2 and new John Deere Category 3 equipment; it let the operator attach integral tools without leaving his seat. The by-now-famous posture seat, power brakes, power steering, wide rear fenders, and dust shields were all standard.

The 5010 was also the first John Deere Tractor available with a 1000-rpm PTO only. Its power was simply too much for a 540-rpm PTO (see PTO discussion under 435, page 25).

Finally, as optional equipment, a new 24.5-32 tire was available. It had been developed jointly by John Deere and tire company engineers to provide additional traction, flotation, and weight-carrying capacity for the 5010.

The 5010 was rated at 121.12 PTO hp and 105.92 drawbar hp. See State of Nebraska Test No. 828 (1962).

The story is told that one of John Deere's early partners found him working on one of his early plows and chided him for time and effort being spent improving what was already the best and most widely sought plow on the market. "We must continue to improve our product," Deere reportedly answered, "or others will and we will lose our trade." In much the same spirit, the 3010 and 4010 were replaced just three years after their introduction. Reason: The engineers had developed something better... the 3020 and 4020, the latter destined to become the most popular tractor of its time. (More about the 4020 on the next page.)

The 3020 and 4020 were more powerful than the tractors they replaced. But the big news was Power Shift transmission. The operator could use a single lever to shift between any of eight forward or four reverse speeds, under load, on the go, and without clutching. Those forward speeds were practical, properly spaced, and ideal for any tractor job. Combined with the flexibility of variable-speed engines, this permitted pinpoint selection of power and speed. An "inching" pedal provided "clutch action" for maneuvering in tight quarters and backing to implements. The time-proved Syncro-Range transmission was available as an option to Power Shift.

The 3020 and 4020 also introduced a hydraulic power differential lock which, when a foot pedal was depressed, directed power equally to both drive wheels for improved traction in mud, snow, or other difficult operating conditions; and a power weight transfer hitch.

The 3020 was originally rated at 65.28 PTO hp. See State of Nebraska Tests No. 848 (1963), No. 851 (1963), No. 852 (1963), No. 940 (1966), No. 941 (1966), No. 942 (1966), No. 1010 (1969) and No. 1011 (1969).

(See next page for 4020 ratings, State of Nebraska Test numbers and dates.)

3020—1964-1972

4020—1964-1972

4020—1964-1972

"The Long Green Line™" of John Deere Tractors has included several that were classic—the "D", the "A", and the "G", among others. But none wears that designation more deservedly than the 4020. It was the most popular tractor of its time. Why? Because it was a tractor a step ahead of its time, and modifications during its lifetime kept it ahead.

Family farming in the 60s was a changing way of life. Old values remained, but the farm came to be looked upon as an "agribusiness". The 4020 was uniquely suited for such farming operations because it, too, was "all business".

It was an era when farmers were doubling up field operations to achieve economies by reducing the number of trips over their fields. The 4020 had the power to accomplish this, pulling not just a disk, or a planter, but both, hooked in tandem, in the earliest steps toward minimum tillage. It pulled not just a baler, but a baler plus a bale ejector plus trailing wagon to receive the bales.

It was an era of greater attention to operator safety features. John Deere engineers developed Roll-Gard®, the first widely accepted tractor roll-over protective structure (ROPS). In 1966 it was offered on the 4020 and other "New Generation" tractors. Then the patents were made available to the entire industry so all tractor operators could have the protection John Deere engineers had made possible.

It was an era when business farming required improved traction for "non-stop" farming. So the engineers developed power front-wheel drive, and made it available first on the 4020. And a better method of attaching dual real wheels was developed to replace the old system of bolting extra wheels to the drive wheels mounted on the tractor's axle. A longer rear axle was offered on the 4020 so the outer wheel could be attached to the axle. To distinguish this system from the previous practice, John Deere called them "double rear wheels".

The 4020 saw two other significant steps in tractor evolution. It was originally available in row-crop or standard configurations with gasoline, diesel, or LP-gas engine (the most popular version was the row-crop diesel shown in the cutaway drawing, right). In the late 60s, demand for LP-gas tractors waned, and that option was dropped. Similarly, the distinctions between standard and row-crop tractors diminished; eventually the two types were blended into one.

The heart of this classic tractor was an equally classic engine with 404-cubic-inch displacement. Improved over the years with many refinements, turbocharging, and intercooling, it remained in John Deere Tractors even after the 4020 had been discontinued.

The 4020 Diesel was originally rated at 91.17 PTO hp. See State of Nebraska Tests No. 849, No. 850, No. 853 (1963); No. 930 (1965); No. 934, No. 939 (1966); No. 1012, No. 1013, and No. 1024 (1969).

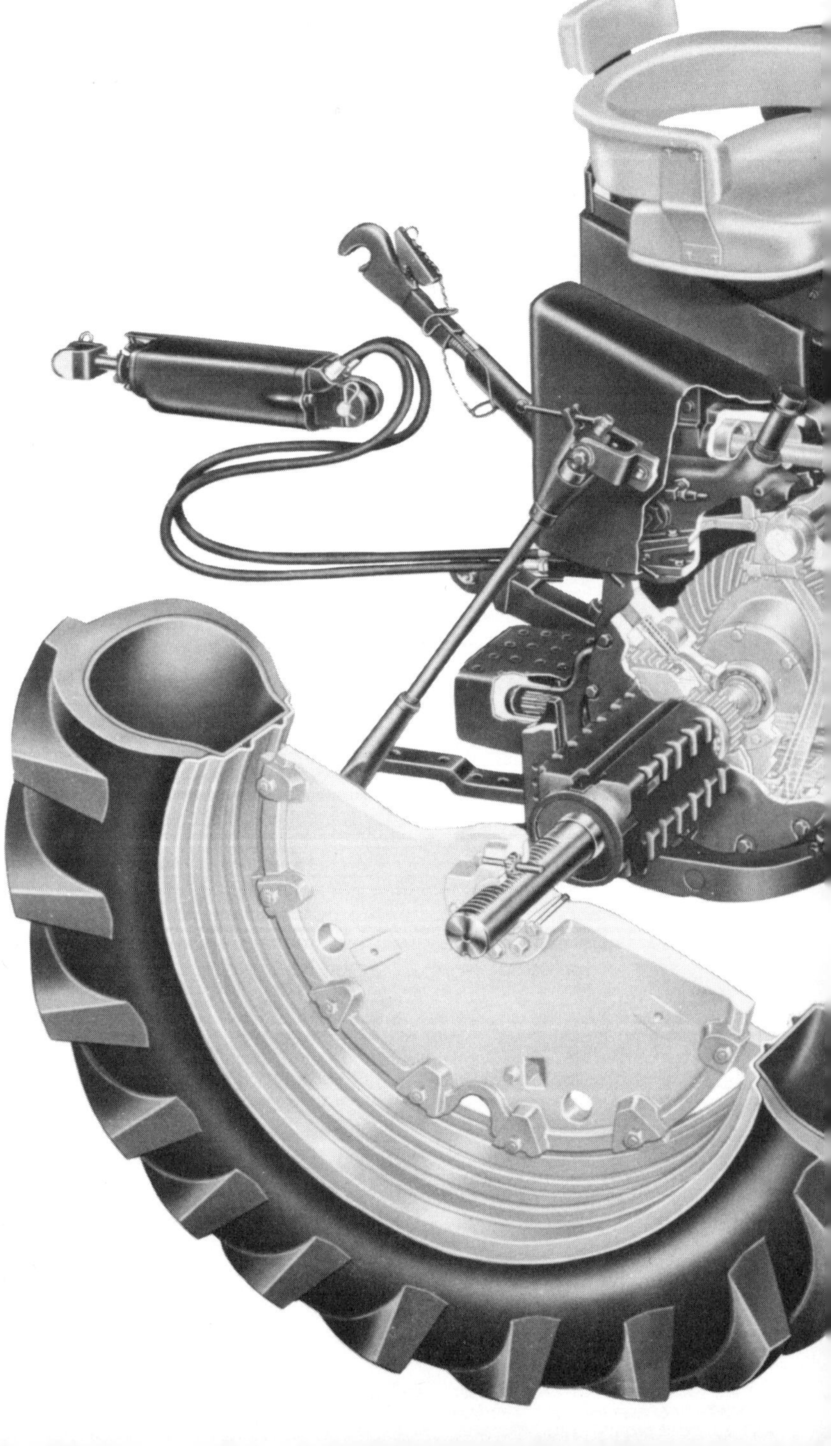

4020
DIESE
ERE

Significant innovations in the big Waterloo tractors invariably inspired demands for more of the same features in the smaller utility tractors. Sometimes this was impractical because of the basic distinctions between the two types of tractors. At other times, the innovations could be adapted to the smaller tractors. The 1020 (below) and 2020 (below, right) provide examples.

These were the first utility tractors with:

Closed-center hydraulics.

3-point hitch with lower-link sensing.

Planetary final drives.

Engage-on-the-go differential lock that automatically disengaged when the difficult field condition was passed.

The 1020 was powered by a 3-cylinder gasoline or diesel engine. Both provided fully governed power over a working range of 1500 to 2500 rpm. The 8-speed transmission, with four gears in a high and a low range, also provided four reverse selections in the low range. Later models offered optional Hi-Lo hydraulic shift to increase pull power, or direction reverser for shuttle shifting.

Both the 1020 and 2020 were offered in three configurations: Low Utility, which could be ordered with a rear exhaust for working in orchards and buildings; Row-Crop Utility; or High Utility. A 1020 Grove Tractor with shielded fenders and low profile especially for use in orchards was also offered.

The 1020 diesel was rated at 38.92 PTO hp. See State of Nebraska Tests No. 937 (1966) and No. 935 (1966).

1020—1965-1973

The 2020 Diesel developed an enviable reputation for its lugging ability. This was due to a remarkable 4-cylinder engine.

Like all John Deere engines, it was built for strength and stamina. Cylinder block and crankcase were a single extra-heavy casting. The crankshaft, a one-piece forging, was statically and dynamically balanced . . . with connecting rod journals overlapping more than 25 percent for strength. Two counterrotating balancer shafts—an extra that John Deere considers vital on 4-cylinder engines but which many competitors ignore—reduced vibration from the inherent imbalance in any 4-cylinder engine.

Forged aluminum-alloy, cam-ground, 3-ring pistons had cups in the crown of the diesel version which swirled the fuel-air mixture for more complete combustion, to get every last ounce of power from each cycle. Direct fuel injection meant faster starts in cold weather and contributed to the torque reserve, which is the basis of lugging ability.

The 2020 and 1020 had a 2-stage, foot-operated clutch that was handy for PTO work. Depressing the pedal part way disengaged the transmission clutch; depressing it all the way disengaged the PTO clutch. Thus, the forward motion of the tractor could be stopped while power-driven equipment digested a slug in unusually heavy crops.

The 2020 diesel was rated at 54.09 PTO hp. See State of Nebraska Test No. 938 (1966); also No. 936 (2020 gasoline, 1966).

2020—1965-1971

2510—1966-1968

Smallest of the "New Generation" row-crop tractors was the 2510, designed for farmers who wanted big-tractor features but didn't need the power of the larger tractors.

Like its bigger brothers, the 2510 was available with a Power Shift or Syncro-Range transmission, and with hydraulics to take the muscle work out of steering, braking, shifting, controlling equipment, actuating the PTO, and engaging the differential lock.

It had the same adjustable posture seat as the larger models.

A 4-cylinder gasoline or diesel engine powered the 2510. That engine, though new to farm tractors, had been five years in development and was thoroughly tested in self-propelled John Deere industrial equipment. It provided governed speeds within a range of 800 to 2500 rpm, and had a working range of 1500 to 2500 rpm. Its features included a heavy-duty crankshaft with five main bearings for strength and rigidity, two balancer shafts, and open-chamber combustion design for faster cold weather starts and a bonus in fuel economy. Forced engine ventilation, positive lubrication of all moving parts, bypass cooling system, and wet-type cylinder sleeves were all features this engine shared with the larger tractor engines.

The 2510 was available with a variety of front-end options—dual-wheel tricycle, regular or heavy-duty Roll-O-Matic, and a regular adjustable or a wide adjustable front axle. A Hi-Crop 2510 also was available.

The 2510 Diesel with Syncro-Range transmission was rated at 54.96 PTO hp. See State of Nebraska Test No. 916 (1965); also No. 913, 914, and 915.

5020—1966-1972

Two years after the 5010 broke the 100 PTO hp barrier for 2-wheel-drive tractors by offering 121 hp, its successor became the first with more than 130 hp. The 5020 scored 133.25 PTO hp in its first test, and 141.34 in a retest.

The 5020 was the most powerful standard tractor of its time, pulling 32-foot tiller-seeders across the Canadian prairies, handling 7-bottom semi-integral plows in Corn Belt soils at speeds up to 5 mph, and throwing up eight beds at a time in cotton country. It proved itself in the mud, water, and abrasive soils of Delta rice fields where lesser tractors met their doom crossing potholes and levees. Its sure-footed stability let it cling to the hillsides of Washington, Oregon and Idaho wheatlands where only crawlers dared go before the 5010 ended their one-time near-monopoly. The 5020 extended the good work the 5010 had begun.

A year after the 5020 Standard was introduced it was joined by the most powerful row-crop tractor of its time. You guessed it—the 5020 Row-Crop, with all the strength and stamina of the standard version but a few additional features of its own. With single or dual rear wheels, it could breeze through open-field tillage operations. Then, equipped with 18.4-38 tires, it could settle comfortably between 30-inch rows to cultivate as many as 12 rows at a time.

The 5020 was rated at 133.25 PTO hp and 116.35 drawbar hp. See State of Nebraska Test No. 947 (1966). See also Test No. 1025 (1969).

820—1968-1973

The 820 (left) was the smallest of the "New Generation" line. Powered by a 3-cylinder diesel engine, it produced 31 PTO hp (factory estimated because the 820 was never tested at Nebraska).

It was a little friend from abroad—manufactured at the John Deere Mannheim Works in Germany, a product of John Deere worldwide design. The 820 proved highly popular during the two short years it was in the line. First, although a low-cost, economy model, it was a lively performer. Second, it was remarkably versatile, providing all the power needed on smaller farms and "backup" power on larger operations. It was at home with a 3-bottom integral plow in many conditions; a variety of haying tools; cultivating; pulling a manure spreader; pushing snow with a blade in the wintertime.

General trends in the 1960s saw farms continuing to grow larger. Their power requirements likewise grew larger, and tractor power increased to meet the demand. But within the general trend, there were exceptions: smaller farmers who chose to maintain their small operations and, when necessary, take off-farm jobs to supplement farming income. These small farmers satisfied their power needs with improved models of utility tractors.

Precise needs for each farm varied. Many found the 820, 1020, or 2020 ideal. For others, the 1020 was too small, the 2020 too big or expensive. John Deere filled the gap in 1968 with the 1520 (right), a marriage of 1020 economy and many 2020 features.

The 1520 diesel was rated at 46.52 PTO hp. See State of Nebraska Tests No. 991 (1968), and No. 1004 (1969).

1520—1968-1973

2520—1969–1973

A number of improvements in Waterloo row-crop tractor engines were announced in 1968. Individually, they seemed minor; cumulatively, their effect was sufficient to permit retesting and higher horsepower ratings for the 3020 and 4020. Some of those improvements: Refinements in piston and ring design and cylinder block and liners; new dry-type air cleaner; 12-volt electrical system with alternator.

The effect was even more significant for the 2510—the power increase was sufficient enough to merit a new designation. Thus was born the 2520.

The 2520 was available with another important John Deere first: The Roll-Gard cab, an operator enclosure built around the steel-frame, Roll-Gard® roll-over protective structure (ROPS).

The 2520 was rated at 61.29 PTO hp. See State of Nebraska Test No. 992 (1968); also No. 993 (1968), No. 1002 (1969) and No. 1003 (1969).

4000—1969-1972

The 4000 was the furthest extension of the high-horsepower-to-weight concept. Powered by the same engine as the 4020, its shipping weight was 7,670 pounds, compared with 8,555 pounds for the 4020. (Both weight figures are for diesel-powered, row-crop models equipped for field use.)

It was an economical tractor for the farmer who wanted to "trade up" to more power but couldn't afford to replace his complete line of tools. For example, rather than trade in his 3020 and 4-bottom plow to buy a 4020 and 5-bottom plow, he could go the 4000 route, keep the 4-bottom plow, but pull it fast enough that in a given period of time he could plow as much total acreage as with a 5-bottom plow at slower speeds. In short, the 4000 was built as a runner rather than a lugger. And, it had the same PTO horsepower as the 4020, for PTO applications.

The 4000 was rated at 96.89 PTO hp. See State of Nebraska Test No. 1023 (1969).

A turbocharger is an air pump. It compresses air and forces it into an engine's combustion chambers. There the denser air can mix with greater amounts of fuel than in nonturbocharged ("naturally aspirated") engines. Result: More complete combustion and more power per cycle, without an increase in the cubic-inch displacement of the engine. The first John Deere Tractor with factory-installed turbocharger was the 4520. But it had more than just a turbocharger. It was "Turbo-Built".

To dissipate the increased heat, a byproduct of increased horsepower, 4520 engines had a new lubricating system with almost twice the capacity of the 4020 lubricating system; a new, bigger radiator and fan; redesigned block for increased flow of coolant; new, bigger capacity air cleaner. Drive train components were strengthened to utilize the greater power and survive its additional stresses.

The 4520 was rated at 122.36 PTO hp. See State of Nebraska Test No. 1014 (1969); also No. 1015 (1969).

4520—1969-1970

4320—1971-1972

The "super 4020". That's what they called this John Deere 4320 with factory-installed turbocharger, introduced in 1970. It was an apt description, for little distinguished the turbocharged newcomer from the more familiar workhorse other than the turbocharger and the 21-percent-greater power it produced. There were a few other differences: Weight was increased by 695 pounds by building heavier engine and transmission components, to keep the extra power from imposing undue stresses.

And, the 4320 had a new tire, the 20.8-34, because the engineers felt no then-existing tire quite matched the 4320 tire requirements. The creation of a new tire to match a new tractor was not unprecedented; it had been done earlier for both the 5010 and the 4520.

The 4320 was rated at 116.55 PTO hp. See State of Nebraska Test No. 1050 (1970).

4620—1971-1972

Turbochargers compress air and force it into the combustion chambers, where the denser air can support the burning of more fuel and thus generate more power with each stroke. But the compression by the turbocharger also heats the air. If the air could be cooled, it would be even more dense, would support still more combustion, and thereby generate even more power. Enter intercoolers.

Intake manifold intercoolers on turbocharged engines cool the air, producing a denser fuel-air mixture which translates into greater power. They are often used on diesel engines for industrial applications and such other uses as over-the-road transport. John Deere engineers were the first to apply their benefits to farm tractors. Result: the 4620, introduced in 1971, with 135.76 PTO hp in official tests.

Consistent with the "Turbo-Built" philosophy, the 4620's engine and transmission were built to handle the high horsepower output. Final drives, axles, and frame were similarly strengthened. And obviously, only a 1000-rpm PTO could handle such power.

Yet the 4620 was no lumbering heavyweight. "Fast on its feet", it handled like much smaller tractors. It adapted efficiently to lighter loads, and could be equipped with single or dual rear wheels to fit requirements of the job at hand. It also had a full complement of "New Generation" features—Power Shift or Syncro-Range transmission, and such hydraulic features as Power Front-Wheel Drive, Power Weight-Transfer Hitch, and power differential lock available as options.

The 4620 was rated at 135.62 PTO hp. See State of Nebraska Tests No. 1064 and No. 1073 (1971).

7020—1971-1975

Paralleling efforts to develop greater tractor power have been efforts to most effectively utilize that power, particularly by increasing traction. One widely adopted technique for increasing traction was adding weight. Another is "dualing", adding an extra set of drive wheels to the rear axle. A third was power-assisted front-wheel drive. Finally, a tractor type that transmitted engine horsepower to all four wheels at all times: The 4-wheel-drive tractor.

The first John Deere "4-wheel-drivers" were the 8010 and 8020. Next, in 1969, came the WA-14 and WA-17. These wheatland giants prefaced the horsepowers to come in the 70s. Factory-observed ratings were 225 hp at the flywheel and 178 hp at the drawbar for the WA-14; 280 flywheel and 220 drawbar hp for the WA-17.

Introduced in 1970 and featuring a John Deere powertrain, the 7020 was unlike any other 4-wheel-drive tractor. It had a wide range of wheel adjustments that fit a variety of row spacings. Tire options included an 18.4-34 size to fit between narrow rows. It had ample under-axle clearance for cultivation; a narrow hood for the visibility row-crop work requires; 1000-rpm, transmission-driven PTO; 3-point hitch with lower-link sensing. In summary, a 4-wheeler for row-crop country as well as the wide-open spaces. In this respect, it was an industry first. One final point: All four wheels could be dualed (as shown above) to further increase traction and provide additional flotation for working soft or muddy ground.

The 7020 was rated at 146.17 PTO hp. See State of Nebraska Test No. 1063 (1971).

The last members of the "New Generation of Power" were introduced in 1971: The 6030 (below), available as a standard or row-crop tractor, and the 7520 (right), a 4-wheel-drive model. Both were powered by a new 6-cylinder, 531-cubic-inch, valve-in-head, variable-speed, turbocharged, intercooled, and "Turbo-Built" engine. Both were designed to stretch one man's working capacity in an era when hired farm help was increasingly scarce and farm operations growing larger.

The 6030, Row-Crop or Standard, with single or dual wheels, fit into the largest and most demanding farm areas of the country. It was equally at home wading through riceland mud, on the rugged slopes of the Pacific Northwest's Palouse Hills wheatfields, in 30-inch corn rows, 40-inch cotton rows, or fighting the calendar to get all the work done on time in northern areas with short growing seasons. Naturally, it was a high-horsepower-to-weight tractor to hustle over large acreages in a hurry, but with lugging ability to "hang in there" pulling big loads in tough ground conditions.

Perhaps the best measure of its capability—and popularity—is the fact that the 6030 remained in The Long Green Line well into the era of Generation II, which you will read about on following pages.

The 6030 was rated at 175.99 PTO hp. For details, see State of Nebraska Test No. 1100 (1972).

6030—1972-1977

The 7020 quickly established a role for 4-wheel-drive tractors on larger operations in row-crop country. It had barely made its debut when zealous new converts set up a familiar chant:

"We like the features... but we need more power!"

This time, the engineers had anticipated the demand—all part of the horsepower race—and early in 1972 the 7520 took its place alongside the 7020.

Externally, the two tractors looked like twins. Internally, there was a 25-hp difference. The 531-cubic-inch engine, which gave the 6030 175-plus PTO hp, developed approximately the same installed in the 7520.

Many 7520 features were identical to those on the 7020 and were described on the previous page. Other features common to both:

A Roll-Gard cab, regular equipment, with heater and/or air conditioner optional; articulated steering; offset exhaust muffler and air intake for better visibility; a 78-gallon fuel tank on each side, both of which could be filled from the ground from one side through a common opening; rack-and-pinion wheel tread adjustment, and a Quik-Coupler hitch for attaching Category 2 and 3 integral equipment without leaving the tractor seat.

And those were just the convenience features. Performance features included:

Syncro-Range transmission with eight forward speeds (16 with optional Hi-Lo); complete lighting package for night work and safer highway transport; wide-swing drawbar, and many others.

The 7520 was rated at 175.82 PTO hp. For details, see State of Nebraska Test No. 1101 (1972).

7520—1972-1975

Generation II

The line of John Deere Tractors known as "Generation II" arrived late in 1972. The four 2-wheel-drive tractors in the new line were distinguished by new styling, dramatic innovations for the comfort and safety of operators, new transmission options, and hydraulically controlled Perma-Clutch™ (a virtually lifetime wet-type clutch).

Time-proved features of the previous line were retained, such as closed-center hydraulics, power brakes and hydrostatic power steering. Options included such familiar favorites as 3-point hitch with lower-link sensing; Power Weight-Transfer Hitch; power front-wheel drive.

Farmers' fuel preferences, particularly for larger tractors, had swung so preponderantly to diesel that only diesel engines were offered.

Until Generation II, only automobiles had bodies. Tractors had cabs... add-on enclosures merely dropped over the operator's platform. With the new line, tractors had bodies, or more precisely, Sound-Gard® body, designed like the automobile body as an integral part of the entire vehicle. A Sound-Gard body might be compared to the command module on a space vehicle. Like the working home of astronauts, Sound-Gard body was a self-contained unit with refined posture seat, and a Control Island housing all instruments, power steering and brake mechanisms, and PTO and clutch controls. The frame of the Sound-Gard body had a sturdy 4-post Roll-Gard structure for roll-over protection.

Equally important were a raft of innovations aimed at reducing sound levels to less than 85 dB(A): curved front and angled sides; sound-absorbing blankets of space-age lead septum; vibration-absorbing rubber body mounts; and more. They succeeded. The 4230, for example, scored 83 dB(A) in official tests with Sound-Gard body.

The combination of comfort and performance features on the "Generation II" tractors inspired a new name: Sound-Idea™ line. The word "sound" as used in Sound-Gard® body implied the noise-reducing ability of the modular body. As used in Sound-Idea, "sound"—synonymous with valid—referred to the entire range of features incorporated into these tractors. The continuing popularity of these tractors proved the validity of the Sound-Idea designation.

The introduction of the "Generation II" line of tractors coincided with an unprecedented rise in net farm income. The demand for these tractors, backed by farmers' ability to buy, so outpaced manufacturing capacity that expansion of the Waterloo factory, and construction of an all-new engine plant there, were undertaken to bring productive capacity closer to the continuing demand.

It is unfortunate that the limitations of this historical look at all John Deere Tractors prohibit closer examination of the many features of the Sound-Idea tractors. Engine details alone would fill a large volume. Perma-Clutch would deserve at least a chapter. Suffice it to say that the "Design in Steel" awards from the American Iron and Steel Institute were bestowed on the entire line collectively.

4030—1973-1977

Smallest of the "Generation II" Tractors was the 4030, powered by a 6-cylinder, 329-cubic-inch, variable-speed, wet-sleeve, valve-in-head engine. Working through an 8-speed Syncro-Range, or new 16-speed Quad-Range™, or 13-speed Creeper transmission, that power made the 4030 a spirited, multipurpose tractor with a broad range of applications. Tricycle or adjustable front axle options contributed to its versatility. The 4030 was rated at 80.33 PTO hp. See State of Nebraska Test No. 1111 (1973).

4230—1973-1977

The 4230 featured all of the important innovations in the line and was available in Row-Crop, Low Profile, and Hi-Crop configurations. This meant versatility, so much so that virtually any farm could equip a 4230 with features ideally suited to its needs. At home in the comfort and quiet of the Sound-Gard body, the operator sensed a new agility and torque reserve. The 4230 was rated at 100.32 PTO hp. For details, see State of Nebraska Test No. 1112 (1973).

4430—1973-1977

Just as the 4020 was the most popular tractor of its time, the 4430 was the most popular tractor of its era. And for the same reason—it was a tractor in step with the times. Its turbocharged engine produced power that placed it in the heart of the most popular tractor power range. The 4430 could be ordered equipped for row-crop or open-country farming, or in a Hi-Crop configuration. A variety of front-end options were available, including Power-Front-Wheel Drive. The 4430 was rated at 125.88 PTO hp. See State of Nebraska Test No. 1110 (1973).

4630—1973-1977

The 4630, largest of the Generation II row-crop tractors, provided all the power needed on many of America's larger-than-average farms and ranches. And, when the owners of smaller Sound-Idea tractors expanded their operations and needed more power, the 4630 satisfied their needs. When these farmers and ranchers did step up to 4630 power, everything was the same except the capacity to work...and that was considerably higher. The 4630 was rated at 150.66 PTO hp. For details, see State of Nebraska Test No. 1113 (1973).

830—1974-1975

In 1973, for the 1974 model year, John Deere introduced a new line of four utility tractors ranging from 35 to 70 hp. But these new tractors were designed for much more than the "chore work" of utility tractors a generation earlier.

The 830 (above) and 1530 (above right) were powered by 3-cylinder diesel engines; the more powerful 2030 and 2630 had 4-cylinder diesel engines. An 8-speed, collar-shift, constant-mesh transmission with 2-lever control and built-in shuttle shift was regular equipment on all four models. So was a 540-rpm PTO.

All had planetary final drives, differential lock, self-adjusting hydraulic wet-disk brakes, and power steering.

The 35-hp 830, smallest of the new line, was available with one or two single-acting remote cylinder outlets or one double-action outlet for its open-center hydraulic system. It was an "all seasons" tractor that paid its way plowing or planting in the springtime; mowing, conditioning, and baling hay all summer; and, with a loader, performing a variety of barnyard maintenance chores any time—even snow removal in the winter. It was a "small" tractor only in relation to the other tractors in "The Long Green Line"; from the standpoint of performance, versatility and dependability, it was a giant.

The 830 was rated at 35.30 PTO hp. For details, see State of Nebraska Test No. 1146 (1973).

1530—1974-1975

A step up from the 830, the 1530 provided additional features that made it even more difficult to categorize this as a "utility" tractor. One of the most important: Hydraulics.

The 1530, as well as the 2030 and 2630, had a closed-center hydraulic system identical in design to those on the larger John Deere Tractors. A variable-displacement hydraulic pump operated only when hydraulic power was in use; horsepower and fuel were conserved when it was not needed.

The closed-center system allowed the hydraulic functions to be used individually or simultaneously, with maximum pressure available for each function. The closed-center system provided instant hydraulic pressure for one or two double-acting remote cylinders; lower-link sensing on the 3-point hitch; engaging the optional independent 540-rpm rear PTO ("live" 540-rpm continuous-running rear PTO was standard): hydraulic brakes; power steering; and optional Hi-Lo transmission.

That Hi-Lo transmission doubled working speeds to 16. The operator could merely flip a lever on the instrument panel and downshift to increase pull power by approximately one third. The increased number of speeds provided by Hi-Lo allowed the operator to select the precise speed best suited to the job at hand, and ease through the tough field conditions that might be encountered.

The 1530 was rated at 45.38 PTO hp. For details, see State of Nebraska Test No. 1147 (1973).

2030—1972-1975

2630—1974-1975

A utility tractor with 26 percent torque reserve? The 2030 had official test reports to prove it, testimony to the engines that powered this and other tractors in the utility lineup. Each of the four tractors had a variable-speed engine designed to handle a wide variety of workloads and job assignments.

In the case of the 2030, the engine was a 4-cylinder, valve-in-head, 4-stroke-cycle type. Diesel, of course. Other features, all shared with the other engines in the line: Built-in strength—cylinder blocks of extra-heavy design, tough enough to serve as a structural part of the tractor. Precision—each block machined to extremely close tolerances so cylinder sleeves would fit snugly and crankshafts revolve without distortion; forged steel crankshafts had a generous overlap between rod and main bearing journals and were statically and dynamically balanced. Coolant surrounded the replaceable wet-sleeve cylinder liners to dissipate heat, protect against hot spots, and ensure longer, more dependable performance.

The 4-cylinder engines on the 2030 and 2630 had balancer shafts running the length of the cylinder block to smooth out vibration. Direct-injection, open-chamber fueling provided fast, easy starts and contributed measurably to the big torque reserve. Dry-type air cleaners filtered dust and dirt from incoming air.

The 2030 was rated at 60.65 PTO hp. See State of Nebraska Tests No. 1085 and No. 1084 (1971).

The 2630 was the most powerful John Deere utility tractor of its time. "Utility" tractor? Yes, in styling and conformation. But no, in the broad area of performance. And its features gave it more in common with the big row-crop tractors of less than a decade earlier. Consider these:

PTO options. While an independent, hydraulically actuated, 540-rpm rear PTO was standard, the 2630 could also be equipped with "live" or independent rear 540/1000-rpm PTO, and 1000-rpm PTO up front.

Its lower-link sensing 3-point hitch could handle Category 2 as well as Category 1 integral equipment. A Quik-Coupler hitch was also available to permit fast, easy hookup of integral equipment without leaving the seat.

Typically, the engineers paid attention to the little things as well as the big things. The dash on the 2630 and other smaller tractors in this line were angled and nonreflective to help eliminate annoying glare. Controls were thoughtfully placed, rockshaft and hydraulic cylinder controls easy to reach at the operator's right. Telescoping lower links on the 3-point hitch made equipment attachment easier. An 8-speed transmission was standard, with Hi-Lo or direction reverser options.

The 2630 was rated at 70.37 PTO hp. For details, see State of Nebraska Test No. 1157 (1974).

8430 (left) and 8630—1975-1978

The 7020 and 7520 represented engineers' concepts of what 4-wheel-drive tractors should be, particularly for row-crop areas. By the mid-70s it was time to give farmers what they wanted next.

Almost as soon as these 4-wheel-drive pioneers made their first tracks across row-crop and open country farmland, the survey teams followed—engineers, sales and marketing executives, all asking: "What do you like about these two 4-wheel-drivers, and what do you want in their successors?" Farmers are seldom reticent when it comes to tractors. They answered. And on the basis of those answers, the 8430 and 8630 were designed and introduced.

The first predictable request: More power. The 8430 provided it; with some 25 more horsepower at the PTO than the 7020, about the same as the 7520.

The second predictable request: Sound-Gard body. John Deere made it standard equipment. It, too, was provided with such additional niceties as an all new Personal-Posture™ seat and a liner for the utility compartment that converted it to a lunch and beverage container.

Third: Adaptability. Rack-and-pinion tread adjustment. A choice of six tire sizes—to match row-crop spacing, to provide traction and flotation to wade through riceland mud, or to hustle across the wide-open spaces of wheatlands.

The 8430 was rated at 178.16 PTO hp. For details, see State of Nebraska Test No. 1179 (1975).

The 8630 answered the demand for more power in dramatic fashion, with 50 more PTO horsepower than the 7520.

The big power of both the 8430 and 8630 came from new 6-cylinder variable-speed, valve-in-head, turbocharged, intercooled and, of course, "Turbo-Built" engines. In the 8430 it was a 466-cubic-inch engine; in the 8630, a 619-cubic-inch engine. But sheer displacement numbers told only half the story. The other half: 20 percent more "torque rise" than the 7520, for lugging power to keep going when the going got tough.

Big and powerful as the 8630 was, it was dimensionally close to the 7520, which meant that this powerhouse was quite at home in the confines of 30-inch rows as well as in open country. Exclusive variable-ratio hydrostatic power steering assisted row-crop work. Both minor steering adjustment while working in narrow rows and turning around at row ends were accomplished with maximum precision and minimum effort.

Periodic servicing was easy on the operator, too. He could drain oil without crawling under the tractor. Both fuel tanks could be filled through one opening, reached from the ground. Oil dipstick, filter and filler were grouped together and easily accessible.

To improve hitch sensitivity and response at the greater draft loads these higher horsepowers could handle, a new all-hydraulic load sensing system was introduced on the 8430 and 8630 hitches.

The 8630 was rated at 225.59 PTO hp. For details, see State of Nebraska Test No. 1180 (1975).

2040—1976-1979

2240—1976-1979

A new line of John Deere Tractors was introduced during the summer of 1975. The most apparent changes from the four replaced models were in the styling and a new Roll-Gard roll-over protective structure. The two smaller models, the 2040 (above) and the 2240 (above, right) were approximately 5 horsepower stronger, respectively, than the 830 and 1530 which they replaced.

In other respects, these tractors were a further step toward erasing the traditional distinctions between row-crop and utility tractors. Such regular-equipment features as power steering, hydraulic brakes, "live" PTO, lower-link-sensing 3-point hitch, engaged-on-the-go differential lock, planetary final drives, and fully adjustable swinging drawbar set these apart from other so-called utility tractors and gave them more in common with their bigger brothers in The Long Green Line. In fact, they were introduced with the slogan: "Family styling that's inherited—family reputation that's earned."

In addition to the four basic models, vineyard and orchard variations of the 2240 were available in narrow, low, compact dimensions for the close work these specialized crop operations required.

The 2040 was rated at 40.86 PTO hp and the 2240 at 50.37 PTO hp. See State of Nebraska Test No. 1191 and 1192 (1976). Horsepowers were the same for the 2030 and 2440, and the 2630 and 2640 (Test No. 1085 and 1157).

2440—1976-1979

2640—1976-1979

2840—1977-1979

About a year after the 4-cylinder 2040, 2240, 2440 and 2640 were introduced, along came the 6-cylinder 2840. The 2840 brought more horsepower in a low-configuration type of tractor.

Its 329-cubic-inch, 6-cylinder diesel engine provided a wide working range of rpm, 1300 to 2500. Support components—cooling, lubrication, air cleaning, fuel delivery—were matched to its power delivery.

The 2840 was equipped with a 12-speed transmission including Hi-Lo as standard. Hi-Lo provided fully-hydraulic, on-the-go shifting through 12 forward speeds and 6 speeds in reverse.

This big "small tractor" was also equipped with closed-center hydraulics, a Category 2 draft-sensing 3-point hitch, and two double-action remote hydraulic outlets, hydraulically-engaged independent 540/1000 rpm PTO, differential lock, integral rack-and-pinion rear axles, and "flat top" row-crop type rear fenders. The battery was located under the raised operator-platform.

The 2840 was rated at 80.65 PTO hp. For details, see State of Nebraska Test No. 1249 (1977).

The Iron Horses

For 1978, John Deere introduced a line of five new row-crop tractors. These tractors had more powerful engines, greatly increased drawbar pull and 3-point hitch lift capacity, and greater hydraulic reserve power than the Generation II Tractors they replaced. In addition, many internal and structural changes were made to back up the power increases. "The new Iron Horses with More Horses and More Iron" were the 4040, 4240, 4240 Hi-Crop, 4440, 4440 Hi-Crop, 4640 and 4840.

All were powered by John Deere diesel engines, naturally aspirated in the 4040 and 4240, turbocharged in the 4440 and both turbocharged and intercooled in the 4640 and 4840. The 4040 had a 404-cubic-inch engine; the other four had 466-cubic-inch displacement engines.

Each tractor in the new line was powered for that tractor's purposes, and bigger, stronger, or heavier internal components were used throughout to provide the stamina to back up the increased horsepower.

Three-point hitch capacities indicate the increased power of these tractors: from 14 to 29 percent more lifting capacity than the hitches on the models replaced. Optional external lift cylinders on the three smaller models increased this power jump further, to as much as 50 percent more lift capacity than their predecessors had. Those external cylinders were standard equipment on the two largest models, as were hydraulic load-sensing hitches for improved draft sensitivity and capacity.

Drawbar performance was enhanced by the more powerful engines and by more iron in the most literal sense; each tractor weighed substantially more than its predecessor. The shipping weight of a new 4640 averaged about 14,350 pounds, while a comparable 4630 weighed only 13,380 pounds. In addition, each model could carry more ballast than its predecessor.

A new hydraulic reservoir next to the closed-center hydraulic pump was a standard feature on all of "The New Iron Horses". And the transmission charge pump, the usual method of providing oil to the hydraulic pump, was 17 percent larger in displacement than the pumps on previous models. This resulted in significantly greater usable hydraulic power.

"The New Iron Horses" also offered more operator comfort. A unique HydraCushioned™ seat suspension system hydraulically reduced the bumps, dips, jerks, and jolts associated with operating tractors on uneven terrain. More sound-deadening materials in the Sound-Gard body further reduced noise levels.

One model, the 4640, tied the all-time record low for official sound tests at maximum available power with a reading of 77.5 dB(A). The previous 77.5 dB(A) score was achieved by a tractor with significantly less power.

The 4040 was initially rated at 90.80 PTO hp; the 4240, 110.94; the 4440, 130.58; the 4640, 156.30; and the 4840, 180.63. See State of Nebraska Test No. 1267, 1266, 1265, 1264 and 1263 (1977). Also see Test No. 1357, 1358, 1359, 1360, 1361 and 1362 (1980).

4440—1978-1982

4040—1978-1982

4640—1978-1982

4840—1978-1982

4240—1978-1982

950—1978-

850—1978-

In early 1978, John Deere introduced the 850 and 950, billed as "Little Big Tractors" because they were small-sized tractors with big tractor features.

Both tractors featured 3-cylinder, wet-sleeve diesel engines that were water-cooled and fuel-injected. They featured 8-speed manual transmissions. These transmissions were geared to provide a wide selection of ground speeds to match a variety of working conditions. There was a creeper speed of just under one mile per hour and a transport speed of nearly 12 mph, plus six well-spaced forward speeds in between and two in reverse. Two-lever control provided four forward speeds in two power ranges.

The John Deere 850 and 950 Tractors featured a differential lock that could be engaged on the go—a valuable feature for work in loose soil, mud, or snow. When employed, this type of differential lock moves a collar to lock the differential as a single unit so power is instantly and uniformly directed to both rear wheels. Because it could be engaged on the go, it required no stopping, no clutching, and no slowing down; and when the troublesome slippage area was passed, the lock would automatically disengage.

The 850 and 950 were equipped with a 540-rpm rear PTO shaft driven by the transmission and featured an overrunning clutch.

A Category 1 three-point hitch allowed quick hookup of integral equipment, and could be adjusted horizontally with sway chains and vertically with lift links. The hitch locked out hydraulically, and a rate-of-drop valve controlled the speed of lowering the hitch.

Front and rear wheel treads were adjustable to suit different jobs. The front tread of the 850 could be set at just over 42 inches or at 50 inches; the rear tread from just over 42 inches to just under 50 inches, with five settings in between.

The 950 had three front tread settings ranging from just under 45 inches to nearly 57 inches. The rear tread could be set in five 2-inch increments from just over 45 inches to a little over 59 inches.

In addition to the tractors, John Deere also introduced power-matched equipment for them; agricultural implements for small-scale farmers who were serious about their acreage and for truck farmers who needed big-tractor features at small-tractor prices. Other implements provided tools for professional groundskeepers, landscape contractors, and nurserymen.

The 850 was rated at 22.27 PTO hp and the 950 at 27.36 PTO hp. See State of Nebraska Test No. 1280 and 1281 (1978).

John Deere introduced two 4WD tractors for 1979, the 8440 and 8640. Compared with the 8430 and 8630 tractors they replaced, the new models offered slightly increased horsepower and numerous internal mechanical changes to meet the increasing demands of bigger acreages, longer days and stretched seasons.

The 8640 featured new pistons, piston rings, piston pins, rods, rod bearings and main bearings. It had new flange-type head bolts with increased load-bearing area. A new turbocharger design improved fuel economy. Both tractors boasted larger radiators with more frontal area and a new dual-pressure system to minimize coolant loss, and new fans and water pumps for quieter 1800-rpm operation. A Perma-Clutch™ was standard equipment.

A common oiling system that joined the front differential with the rest of the cooled and filtered hydraulic oil system eliminated separate oil for servicing of the front differential.

Powertrain improvements included increased cooling capacities; new helical-cut Hi-Lo planetary gears; new 4-gear drop train; new transmission gears, bearings and synchronizers; and new final drive parts including 100 mm drive axles for both models.

Such options as hydraulic front and rear differential locks, rockshaft with increased lift capacity, and the choice of up to four remote cylinder outlets were offered to increase field performance and reliability.

Four-wheel drive tractors were being worked considerably longer hours now than just a few years earlier, so several changes were made to make the operator more comfortable. A new HydraCushioned™ seat suspension system with hydraulic verticle suspension and spring-dampened lateral suspension automatically smoothed out jolts. A horizontal shock absorber reduced the side-to-side motion characteristic of articulated 4-wheel-drive tractors.

One lever controlled the 16-speed Quad-Range transmission for easy shifting to the ideal speed for the operation. An engaging pawl eased and smoothed engagement or disengagement of the built-in park lock.

A new Investigator™ warning system monitored many critical engine and powertrain functions, and warned of malfunctions both audibly and with indicator lights. A built-in computer logic system protected against the "false alarms" that could be triggered in a less sophisticated system.

The 8440 was rated at 179.83 PTO hp and the 8640 at 228.75 PTO hp. See State of Nebraska Test No. 1323 and 1324 (1979).

8440—1979-1982

8640—1979-1982

For the 1980 model year, John Deere tractors in the 40- to 80-horsepower category were restyled to match the modern look of over-90-hp John Deere models. Also, front-wheel-drive assist became available for the first time on John Deere tractors of this size.

Mechanical front wheel drive was offered on the 2040, 2240, and the new 2940 which replaced the 2840. In soft fields, mud or snow, tractors equipped with this option had pulling power that could rival 2-wheel-drive tractors with up to 20 percent more rated horsepower.

The integrated design of the front-wheel-drive components provided good front-axle clearance, at the same time reducing the need for front and rear weighting to maintain balanced weight distribution. The front-wheel drive could be engaged or disengaged on the go, and was equipped with a limited-slip front differential that could be engaged as needed, then would disengage automatically when the need was past.

The 2940 featured a new top-shaft-synchronized transmission package with Hi-Lo shift, providing 16 forward and 8 reverse speeds. On the 2040 and 2240, an 8-speed top-shaft-synchronized (TSS) transmission was packaged with the optional mechanical front-wheel drive.

These tractors were powered by 3-, 4-, or 6-cylinder diesel engines, as were their predecessors, with such field-proven features as direct fuel injection, wet-sleeve cylinder liners, bypass cooling, and pressure lubrication of all vital parts. Additionally, these engines were improved with beefed-up oil pump, improved lubrication of critical bearings, and specially-coated piston rings to improve service life.

The improvements increased the reliability of these engines and powertrains so much that John Deere warranted the engine block, head, timing cover, pan, rocker arm cover, and all parts within . . . and the clutch housings, transmission case, final drive housing, and all parts enclosed within . . . as well as internal components of the mechanical front-wheel drive, for two years or 1500 hours, whichever came first. This was in addition to the regular John Deere agricultural equipment warranty.

Three-point hitch rockshaft capacity was increased by more than 20 percent on the 2440 and 2640. Both had longer and 50 percent stronger telescoping draft links, as well.

Operators of these John Deere 40- to 80-hp tractors found that the new seat was adjustable for individual height, weight, and reach. The seat could be tilted back five degrees for a more stable ride, and it was standard on all models except orchard/vineyard tractors.

The Sound-Gard body popularized on higher horsepower tractors was made available for the 2940, too.

The 2040 was rated at 41.25 PTO hp, the 2240 at 50.90 and the 2940 at 81.17. See State of Nebraska Test No. 1350 and 1351 (1980), and 1393 and 1394 (1981). Horsepowers were not changed for the 2440 and 2640. (Refer to Test No. 1085 and 1157.)

2040—1980-1982

2240—1980-1982 Orchard/Vineyard version shown

2440—1980-1982

2640—1980-1983

2940—1980-1982

8650—1982- 8850—1982- 8450—1982-

In the early 1980s, expectations continued to rise of what could be accomplished using four-wheel-drive tractors. More horsepower was needed, in ever more refined form, to reduce manpower costs by increasing productivity. Thus came the new 8450, 8650 and 8850.

The 955-cubic-inch John Deere V-8 engine which powered the 8850 made this the most powerful John Deere tractor ever produced. It put John Deere into the 300-PTO-hp power size. The 8450 and 8650 also offered more power than the models they replaced.

The 8850's new Category 4 hitch was convertible to 4N, with 18 percent more lift capacity than the hitches on earlier John Deere 4-wheel-drive tractors. Two lift-assist cylinders achieved this increased capacity. Dual sensing cylinders provided the same hydraulic draft-link sensing capabilities as the single-cylinder system used on 8450 and 8650 models.

The main hydraulic pump on the 8850 provided 50 percent more capacity than the pump on the 8640. New selective control valves with ISO couplers were used on all three models. The larger openings in these valves allowed increased hydraulic flow, thus providing more power for such operations as orbital motors and hydraulic pumps, and permitting implements to be raised and lowered quickly for faster turnaround in shorter headlands.

Engine refinements that reduced parasitic power losses gave this new line excellent fuel economy. For example, the 8450 and 8650 were equipped with mixed-flow fans to move cooling air in both radial and axial directions. These fans required less power, but did not reduce cooling capacity. In some test conditions this design saved as much as 5.4 hp.

The 8850 was equipped with temperature-sensing viscous fan drive to adjust the fan speed automatically; when temperatures dropped, fan speed was reduced, again freeing power for productive work.

Within the engines, too, refinements to reduce friction cut parasitic power losses, with the same result: more work from the same amount of fuel. Improved turbochargers and a higher precision matching of fuel injection pumps likewise reduced parasitic power loss.

Because the refined engines on these tractors performed so well at lower engine speeds—2100 rpm as opposed to the 2600 to 2700 rpm range on some 4-wheel-drive tractors—less stress was imposed on engine components. Longer engine life was achieved.

The Investigator warning system introduced on John Deere 40 Series 4-wheel-drive tractors was improved with the Investigator II system, standard equipment on the new 50 Series. Investigator II included a new microprocessor for faster, more accurate response and more information.

The 8850 was rated at 303.99 PTO hp, the 8650 at 238.56 and the 8450 at 186.98. See State of Nebraska Test No. 1434, 1435 and 1436 (1982).

At a glance the 1050 looked like a carbon copy of the 850 and 950 Compact Utility Tractors introduced two years earlier. In fact, these three models were very similar.

However, the 1050 was significantly different in key ways. For example, its 104.8 cubic inch 3-cylinder diesel engine was turbocharged. Of course, turbocharging offered greater power in a smaller package. When compared with the 950's engine of the same displacement, the 1050 engine delivered nearly 23 percent more PTO horsepower. In addition, this extra power was delivered with less engine noise because the turbocharger was working in the exhaust system, acting as a muffler and absorbing engine noise.

Additional horsepower increased work capability, so the 1050 was equipped with a load-sensing 3-point hitch. This mechanical center-link sensing system increased control of working depth in a variety of soil conditions. Also, the continuous-running 540-rpm PTO increased productivity when working with PTO-driven equipment.

The 1050 was rated at 33.41 PTO hp. For details, see State of Nebraska Test No. 1327 (1979).

Two more tractors joined the John Deere Compact Utility line for 1981 . . . the 650 and 750. Liquid-cooled diesel engines powered both models, the 650 engine had two cylinders, the 750 had three cylinders. They featured a sliding-gear transmission with eight forward and two reverse speeds. A 540-rpm PTO, Category 1 three-point-hitch, and adjustable drawbar were standard equipment. Mechanical front-wheel-drive, power steering, and front PTO were optional.

Both models featured a 6.2-gallon fuel tank, dual brake pedals, 12-volt starting and lighting system, mechanical differential lock, and a cold-weather starting aid. The muffler could be converted from horizontal to vertical position by removing a few bolts and repositioning it.

These mini tractors were advertised at 17 and 20 gross engine horsepower, for the 650 and 750 respectively. The 650 was not tested at the Nebraska Test station. The 750 was tested in 1983 and rated at 18.54 PTO hp (see Test No. 1504).

750—1981-

1050—1980-

650—1981-

1250—1982-

1450—1984-

At the time they were tested at the Nebraska Test center, the 1250, 1450, and 1650 Tractors set nine records for fuel efficiency and lugging ability. The 1650 was the most fuel-efficient tractor ever tested, regardless of horsepower. And compared with other tractors in the 55- to 65-hp size, the 1650 set records in four fuel-efficiency categories: PTO at rated engine speed, PTO at standard speed, drawbar at full load, and PTO at varying power and fuel consumption.

The 1450 ranked number one among "currently available" tractors in the 45- to 55-hp class, in PTO fuel economy at rated engine speed and at rated PTO speed, and in drawbar fuel economy at full load. It also offered the best lugging ability for tractors of its size.

In the category for 35- to 45-hp tractors, the 1250 offered the greatest lugging ability at 80 percent of rated engine rpm among tractors available at the time of the test.

The 1250 was powered by a naturally-aspirated 143 cu. in. 3-cylinder diesel engine, while a 190 cu. in. 4-cylinder diesel powered the 1450 (naturally aspirated) and 1650 (turbocharged). All three models had 3-range sliding-gear collar-shift transmissions giving nine forward and three reverse speeds.

The 1250 was rated at 40.70 PTO hp, the 1450 at 51.39 PTO hp, and the 1650 at 62.22 PTO hp. For details on these and the fuel efficiency test results, see State of Nebraska Test No. 1437 (1982) and 1505 and 1506 (1983).

1650—1984-

The Longest New Line

Late in 1982, John Deere unveiled the largest number of new models the company had ever introduced at one time. These were the "50 Series" 2-wheel-drive tractors, 10 new models in the 45 to 190 hp range.

Glancing at the differences in horsepower tells the story: they represented a complete range of tractors with well-spaced horsepowers. The 2150 was advertised at 45 PTO hp, the 2350 at 55 hp, the 2550 at 65 hp, the 2750 at 75 hp, the 2950 at 85 hp, the 4050 at 100 hp, the 4250 at 120 hp, the 4450 at 140 hp, the 4650 at 165 hp, and the 4850 at 190 hp.

These tractors offered significantly improved fuel economy and productivity, exactly what farmers needed to help combat the huge jump in fuel prices and the drop in commodity prices which they were experiencing at that time.

Fuel efficiency had become critically important. In 1970, according to government surveys, fuel represented 25 percent of the cost of operating a tractor. By 1980, this had increased to 35 percent. Much of the fuel economy in this 50 Series resulted from a combination of mechanical front-wheel drive and—for the five largest models—a new 15-speed Power Shift Transmission. But other improvements (varying by model) helped, too: a temperature-modulated fan drive that could speed up or slow down, depending on engine cooling requirements; turbochargers designed to help burn fuel more efficiently at varying loads; low friction pistons; new sealing rings which reduced parasitic power loss in the main clutch of the Power Shift transmission; refinements to the fuel injection system; and a number of transmission and hydraulic system improvements.

Mechanical front-wheel drive allowed up to 40 percent of the tractor's power to go to the front wheels to improve traction, reduce wheel slippage, and improve fuel economy.

Poor maneuverability and turning radius were always serious drawbacks to front-wheel-drive tractors in the past. But this front-wheel-drive option had something to set it apart—a patented new feature called Caster-Action™ front-wheel drive. The front wheels could caster 13 degrees and tilt to the side to clear the tractor's frame. This allowed the front wheels to turn 50 degrees, giving maneuverability similar to that of regular two-wheel-drive tractors.

The 4050, 4250, 4450, 4650 and 4850 could be equipped with a new 15-speed Power Shift transmission, representing the industry's furthest advancement in effortless, on-the-go shifting. As the tractor moved through the field, it could be shifted from gear to gear, with one finger and without clutching. Seven well-spaced speeds fell within the field working range, four speeds were for PTO work under three miles per hour, and four for transport. PTO engagement was modulated in order to take on loads gradually, thus reducing shock loads on shafts and other components. This not only significantly improved reliability, but also improved ease of operation. In addition, modulation made shifting from neutral to any operating speed virtually effortless.

The Power Shift transmission was standard on the 4850. John Deere's Quad-Range transmission was the standard on the next four models in this new line. On the smaller 2150, 2350, 2550, 2750 and 2950, a new top-shaft-synchronized transmission (TSS) was standard. It permitted on-the-go shifting within gear ranges, with little effort. This ability to make fast, easy speed changes meant tractor speeds could be favorably matched to job requirements for most efficient use of power and fuel.

An electronic warning system, on the five largest models, monitored 10 critical vehicle and engine functions and gave the operator both visual and audible signals when something needed attention.

The closed-center hydraulic system pioneered on John Deere tractors more than 20 years earlier, and steadily improved over the years, was further improved on these new models. New selective control valves reduced restrictions in hydraulic fluid flow, permitting faster lifting of heavy loads. New breakaway hydraulic couplers featured improved drip-free coupling and uncoupling under pressure. New rotating dust covers with extensive sealing provided better weather and dust protection for the couplers.

The John Deere Sound-Gard body was offered on models in the 2350, 2550 and 2750 sizes for the first time.

When the 50 Series Tractors were tested in the Nebraska test program, they set several performance records for fuel efficiency, low noise levels inside the operator's enclosures, and drawbar pull.

The largest of the line, the 4850, achieved two of the records. In drawbar pull testing, the 4850 pulled 20,810 pounds, slightly more than 2,000 pounds greater than the previous record of 18,777 pounds. The 4650 pulled 18,794 pounds, also passing the previous record.

In fuel efficiency testing, the 4850 established itself as the most fuel efficient of all 2-wheel-drive tractors over 70 PTO hp then on the market. During PTO rated speed testing, the 4850 produced 17.47 horsepower hours per gallon of fuel. During PTO variable load testing, it produced 14.50 horsepower hours per gallon. Both of these were new records for over-70 hp tractors in production at that time.

The 4050 earned honors as the quietest tractor ever tested in Nebraska. At 50 percent load, reduced engine speed, sound levels inside the Sound-Gard body on the 4050 were recorded at 70.0 dB(A). This broke the previous record of 71.5 dB(A) set just days earlier by the 4250 and was a full 3 dB(A) below any tractor tested prior to the new John Deere tractors. (See State of Nebraska Test No. 1457 and 1458.)

The 4050 also set a new record for sound levels at full load. The sound level was recorded at 73.5 dB(A), quieter than any other tractor ever measured under full load.

The 2150 was rated at 46.47 PTO hp, the 2350 at 56.18 hp, the 2550 at 65.94 hp, the 2750 at 75.35 hp and the 2950 at 85.37. With Power Shift transmission, the 4050 was rated at 100.95, the 4250 at 123.32, the 4450 at 140.43, the 4650 at 165.52 and the 4850 at 192.99. See State of Nebraska Test No. 1457 through 1461 (1982) and 1469 through 1477 (1983). Beginning with 1986 models, the 4050 had a turbocharged engine and was rated at 105.89 PTO hp. See Test No. 1590 and 1591 (1986).

4250—1983- **4450—1983-** **4650—1983-** **4850—1983-**

The 50-hp 2255 Orchard/Vineyard Tractor was specially designed for work in tight spaces. Equipped with standard tires for orchard work, the 2255 measured just over 52 inches to the top of the steering wheel. Equipped for work in vineyards, the 2255 was just 59 inches wide overall. Other specialty versions were available in this line, too... the 2150, 2350 and 2550 could be equipped low, slow, narrow, or wide; the 2750 as either a Low-Profile or Hi-Clearance tractor; the 4250 as a Hi-Crop model. **2255—1983-1986**

855—1986-

755—1986-

655—1986-

Advances such as Sound-Gard bodies, Personal-Posture armchair seats, and synchronized and Power Shift transmissions had over the years made operating larger tractors more and more like operating an automobile or pickup. In 1986, John Deere extended car-like handling to Compact Utility Tractors by introducing the 655, 755 and 855. Hydrostatic drive with 2-pedal speed and direction control, power steering, and cruise control were standard on these models.

A system of pins and brackets let the operator drive into attachments or slide them into place, without tools. A cluster of quick-couplers permitted plugging in hydraulic powered equipment near where attachments locked on.

These tractors featured a 540-rpm rear PTO and 2100-rpm mid-PTO. Implements could be engaged on the go because both PTOs were independent of the transmission. A hydraulic clutch activated the PTO smoothly. Separate PTOs meant a tiller could be operated independently of the mower. An optional mid-mount rockshaft permitted raising or lowering a 3-point hitch too without disturbing the mid-mounted mower. Up to 8 inches of mower-deck clearance permitted loader, tiller and other work with the mower in place.

Individual inboard-mounted wet-disk rear brakes, unitized mainframe, planetary final drives, locking differential, Category 1 three-point-hitch, position-control rockshaft, and optional mechanical front-wheel drive that could be engaged on the go greatly increased versatility and separated these models from similar size machines capable of performing only lawn and garden chores.

Each tractor was powered by a 3-cylinder liquid-cooled engine. Displacement of the 655 engine was 40.2 cu. in., 53.6 cu. in. in the 755, and 60.7 cu. in. in the 855. Engine horsepowers: 16 for the 655, 20 for the 755, and 24 for the 855.

The 900 HC was designed for growers who needed precise control when cultivating between rows of tender young plants in such operations as truck gardens, nurseries, strawberry fields and tobacco fields. Sitting to the right of the engine compartment, the operator view was superb. So was crop clearance: 23 inches under the front axle, nearly 27 inches under the rear axle housing.

The base machine included a cultivator lift system and rear toolbar. Cultivator adjustments included lift height, leveling of cultivator frames, downpressure for mid-cultivator, depth stop, and adjustments for the location of cultivator shanks.

The standard continuous "live" PTO allowed the operator to unhook the rear toolbar and hook up a rotary cutter or tiller to use the 900 HC in other applications.

The muffler was located under the engine compartment hood to provide better forward visibility, as well as reducing sound levels.

Standard features included an 8-speed transmission with two reverse speeds, continuous "live" 540-rpm PTO, differential lock, rear rockshaft and category 1 three-point-hitch with lateral float, mid-hydraulic system with controls, and adjustable front axle.

The 900 HC was not tested at the Nebraska Test center. Advertised horsepower was 25 SAE gross engine hp.

900HC—1986-

3150—1985-

The first John Deere Tractor with mechanical front-wheel drive as standard equipment was the 3150. Not only was this tractor available only with MFWD, but the front-wheel drive engaged automatically whenever high drawbar pull was required. The operator could also engage and disengage the MFWD independent of the automatic system when not under heavy load, in such operations as transporting equipment.

The 3150 featured a 359 cubic inch, naturally-aspirated, 6-cylinder John Deere diesel engine. Matched with a top-shaft-synchronized Hi-Lo transmission, the 3150 offered 16 properly spaced speeds for all field and farmstead operations. Hi-Lo gave the operator no-clutch, shift-on-the-go capability under load.

A 6300 pound hitch lift capacity and superior lugging ability of the engine provided plenty of power to handle 12-row equipment with ease.

The 3150 could be equipped with the optional Sound-Gard body, the time-honored John Deere operator enclosure representing the ultimate in tractor operator comfort. On open operator-station models, controls were conveniently located in a console to the right of the personal contour seat. Tilt-telescoping steering wheel and full instrumentation were standard, along with hydrostatic steering, differential lock, independent "live" 540/1000 rpm PTO, hydraulic disk brakes and 12-volt electrical system.

The 3150 was rated at 96.06 PTO hp. For details, see State of Nebraska Test No. 1589 (1986).

While American farmers continued facing extreme difficulties in their efforts at making a profit, John Deere engineers kept on refining, improving, and designing tractors that were increasingly more fuel efficient and productive. The 2155, 2355, 2555, 2755 and 2955 introduced early in 1987 were examples of these design efforts.

When equipped with synchronized transmissions and lower-speed engines, the new models offered up to 10 percent greater fuel economy than their predecessors. This reduced fuel consumption resulted from a combination of refinements, including: a slower, 2300-rpm rated operating speed, a viscous fan that ran only when needed, and repositioning rings higher up on pistons to reduce compression losses.

A 179 cu. in. naturally-aspirated 3-cylinder diesel engine powered the 2155. Another naturally-aspirated engine, this time 4-cylinder with 239 cu. in. displacement, powered the 2355 and also the 2555 with collar-shift transmission.

The 2555s equipped with top-shaft-synchronized (TSS) transmissions used a turbocharged engine. So did the 2755. The largest model of this series, the 2955, had a 6-cylinder 359 cu. in. naturally-aspirated engine.

Putting the efficient power of these tractors to work in all types of operations and conditions was readily possible because of the wide selection of transmission alternatives available. The 2155, 2355 and 2555 could be equipped with either a non-synchronized collar-shift (CS) or synchronized TSS transmission. Those tractors with the CS transmission could also be equipped with a direction reverser. This permitted the operator to reverse direction hydraulically without clutching, merely by moving a dash-mounted control lever.

The TSS transmission was standard on 2755 and 2955 Tractors. A Hi-Lo hydraulic-shift option doubled the number of gears, to 16 forward and eight reverse, on any tractor with the TSS transmission. (Hi-LO was standard on the 2955.) Also, for work requiring super-slow travel speeds, such as pack-and-carry operations, a TSS creeper transmission was available for the 2355, 2555, 2755 and 2955.

Specially designed for use in orchards and vineyards, two narrow, low-built models were available in this series. They were identified as the 2355N and 2855N.

The 2755 was rated at 76.57 PTO hp and the 2955 at 86.22 PTO hp. See State of Nebraska Tests No. 1605 and 1606 (1986). At the time of this printing, the other models were not yet officially tested, but were advertised with these PTO horsepowers: 2155, 45 hp; 2355 and 2355N, 55 hp; 2555, 65 hp; 2855, 80 hp.

2355N—1987-

2155—1987-

2355—1987-

2555—1987-

2855N—1987-

2755—1987-

2955—1987-

The following serial numbers provide a means of determining a tractor's age. The number opposite each year designation is the beginning serial number for that year for the model designated in the headline. Years are product years (usually August 1 of the preceding calendar year to the following July 31) unless noted as follows: *Calendar Year; **Fiscal Year (November 1 to October 31). For convenience, tractors are listed alphabetically and numerically, rather then chronologically.

"A"

1934	410 000
1935	412 869
1936	424 025
1937	442 151
1938	466 787
1939	477 000
1940	488 000
1941	499 000
1942	514 127
1943	523 133
1944	528 778
1945	548 352
1946	555 334
1947	578 516
1948	594 433
1949	620 843
1950	648 000
1951	667 390
1952	689 880

"AO" and "AR"

1936	250 000
1937	253 521
1938	255 416
1939	257 004
1940	258 045
1941	260 000
1942	261 558
1943	262 243
1944	263 223
1945	264 738
1946	265 870
1947	267 082
1948	268 877
1949	270 646
1950	272 985
1951	276 078
1952	279 770
1953	282 551

"AO" (Styled)

1937	AO-1000
1938	AO-1539
1939	AO-1725
1940	AO-1801

"B"

1935	1 000
1936	12 012
1937	27 389
1938	46 175
1939	60 000
1940	81 600
1941	96 000
1942	126 345
1943	143 420
1944	152 862
1945	173 179
1946	183 673
1947	199 744
1948	209 295
1949	237 346
1950	258 205
1951	276 557
1952	299 175

"BO" and "BR"

1936	325 000
1937	326 655
1938	328 111
1939	329 000
1940	330 633
1941	332 039
1942	332 427
1943	332 780
1944	333 156
1945	334 219
1946	335 641
1947	336 746

"BO" (Lindeman) Crawler

1943	332 901
1944	333 110
1945	333 666
1946	335 361
1947	336 441

"D"

1924	30 401
1925	31 280
1926	35 309
1927	43 410
1928	54 554
1929	71 561
1930	95 367
1931	109 944

1932	115 477
1933	115 665
1934	116 273
1935	119 100
1936	125 430
1937	130 700
1938	138 413
1939	143 800
1940	146 566
1941	149 500
1942	152 840
1943	155 005
1944	155 426
1945	159 888
1946	162 598
1947	167 250
1948	174 879
1949	183 516
1950	188 420
1951	189 701
1952	191 180
1953	191 439

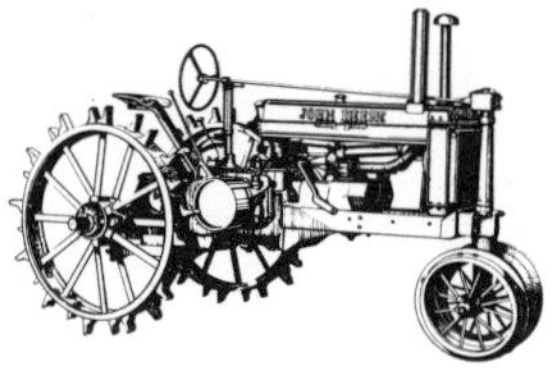

"G"/"GM"/"G"

1938	1 000
1939	7 734
1940	9 321
1941	10 489
1942	12 069
1943	12 941
1944	13 748
1945	13 905
1946	16 694
1947	20 527
1948	28 127
1949	34 587
1950	40 761
1951	47 194
1952	56 510
1953	63 489

"GP"Standard

1928	200 111
1929	202 566
1930	216 139
1931	224 321
1932	228 666
1933	229 051
1934	229 216
1935	230 515

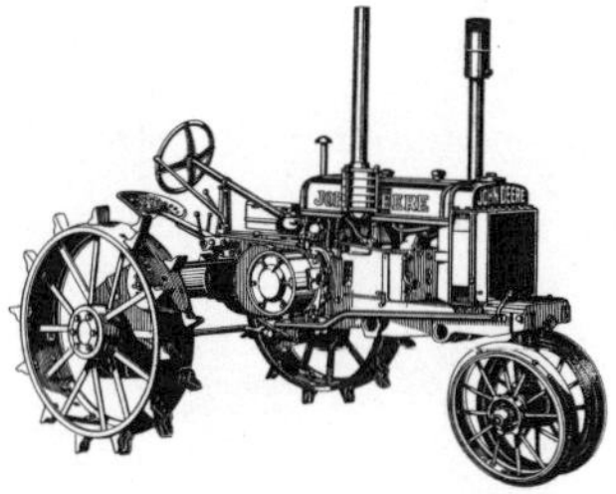

"GP"—Wide Tread

1929	400 000
1930	400 936
1931	402 741
1932	404 770
1933	405 110

"GP-O"**

1931	15 000
1932	15 226
1933	15 387
1934	15 412
1935	15 589

"H"

1939	1 000
1940	10 780
1941	23 654
1942	40 995
1943	44 755
1944	47 796
1945	48 392
1946	55 956
1947	60 107

"L"

1937	621 000
1938	621 079
1939	626 265
1940	630 160
1941	634 191
1942	640 000
1943	640 738
1944	641 038
1945	641 538
1946	641 958

"LA"

1941	1 001
1942	5 361
1943	6 029
1944	6 159
1945	9 732
1946	11 529

"M"

1947	10 001
1948	13 734
1949	25 604
1950	35 659
1951	43 525
1952	50 580

"MC" Crawler

1949	10 001
1950	11 630
1951	13 630
1952	16 309

"MT"

1949	10 001
1950	18 544
1951	26 203
1952	35 845

"R"

1949	1 000
1950	3 541
1951	6 368
1952	9 293
1953	15 720
1954	19 485

Waterloo Boy "L" & "LA"

1914	1 000

Waterloo Boy "N"

1917	10 020
1918	10 221
1919	13 461
1920	18 924
1921	27 026
1922	27 812
1923	28 119
1924	29 520

Waterloo Boy "R"

1915	1 026
1916	1 401
1917	3 556
1918	6 982
1919	9 056

40 Hi-Crop

1954	60 001
1955	60 060

40 Special

1955	60 001

40 Standard

1953	60 001
1954	67 359
1955	69 474

40 Tricycle

1953	60 001
1954	72 167
1955**	75 531

40 2-Row Utility

1955	60 001

40 Utility

1953	60 001
1954	60 202
1955	63 140

"40C" Crawler

1953	60 001
1954	63 358
1955	66 894

50

1952	5 000 001
1953	5 001 254
1954	5 016 041
1955	5 021 977
1956	5 030 600

60

1952	6 000 001
1953	6 007 694
1954	6 027 995
1955	6 042 500
1956	6 057 650

70

1953	7 000 001
1954	7 005 692
1955	7 017 501
1956	7 034 950

80

1955	8 000 001
1956	8 000 755

320

1956	320 001
1957	321 220
1958*	325 127

330

1958	330 001
1959	330 171
1960	330 935

420

1956	80 001
1957	107 813
1958*	127 782

420C Crawler

1956	80 001
1957	107 813
1958*	127 782

430

1958	140 001
1959	142 671
1960	158 632

430C Crawler

1958	140 001
1959	142 671
1960	158 632

435

1959	435 001
1960	437 655

520

1956	5 200 000
1957	5 202 982
1958	5 209 029

530

1958	5 300 000
1959	5 301 671
1960	5 307 749

620

1956	6 200 000
1957	6 203 778
1958	6 215 048

630

1958	6 300 000
1959	6 302 749
1960	6 314 381

650

1981	1 000
1982	3 539
1983	6 250
1984	10 543
1985	15 001
1986	19 001
1987	22 501

655

1986	360 001
1987	420 001

720

1956	7 200 000
1957	7 203 420
1958	7 217 368

730

1958	7 300 000
1959	7 303 761
1960	7 322 075
1961	7 328 801

750

1981	1 000
1982	3 448
1983	5 613
1984	8 597
1985	13 001
1986	18 501
1987	22 601

755

1986	360 001
1987	420 001

820 (2-Cyl.)

1956	8 000 000
1957	8 200 565
1958	8 203 850

820 (3-Cyl.)

1968	10 000
1969	23 100
1970	36 000
1971	54 000
1972	71 850
1973	90 200

830 (2-Cyl.)

1958	8 300 000
1959	8 300 727
1960	8 305 301
1961	8 306 892

830 (3-Cyl.)

1974	108 507
1975	155 914

850

1978	1 024
1979	3 859
1980	7 389
1981	11 338
1982	12 481
1983	14 183
1984	16 006
1985	18 001
1986	22 001
1987	25 501

855

1986	360 001
1987	420 001

900 HC

1986	1 001
1987	1 701

950

1978	1 024
1979	5 229
1980	10 453
1981	14 893
1982	16 204
1983	18 204
1984	20 007
1985	23 001
1986	26 001
1987	28 501

1010

1961	10 001
1962	23 630
1963	32 188
1964	43 900
1965	53 722

1010C Crawler

1960	10 001
1961	13 692
1962	23 630
1963	32 188
1964	43 900
1965	52 722

1020**

1965	14 501
1966	14 682
1067	42 715
1968	65 184
1969	82 409
1970	102 039
1971	117 500
1972	134 700
1973	157 109

1050

1980	1 000
1981	5 280
1982	6 572
1983	9 001
1984	11 006
1985	14 001
1986	17 001
1987	19 501

1250

1982	1 001
1983	1 256
1984	3 001
1985	4 001
1986	5 001
1987	5 501

1450

1984	1 020
1985	2 201
1986	3 001
1987	3 501

1520**

1968	76 112
1969	82 405
1970	102 061
1971	117 500
1972	134 700
1973	157 109

1530

1974	108 811L
	176 601T
1975	145 500L

1650

1984	1 021
1985	2 401
1986	3 001
1987	3 501

2010

1961	10 001
1962	21 087
1963	31 250
1964	44 036
1965	58 186

2010C Crawler

1960	10 001
1961	10 999
1962	21 087
1963	31 250
1964	44 036
1965	58 186

2020**

1965	14 502
1966	14 680
1967	42 721
1968	65 176
1969	82 404
1970	102 032
1971	117 500

2030

1972	134 700T
1973	157 109T
1974	140 000L
	187 301T
1975	145 500L
	213 350T

2040

1976	179 963
1977	221 555
1978	266 057
1979	304 165
1980	336 935
1981	392 026
1982	419 145

2150

1983	433 467
1984	505 001
1985	532 000
1986	562 001

2155

1987	600 000

2240

1976	179 298
1977	221 716
1978	277 267
1979	305 307
1980	337 767
1981	392 292
1982	418 608

2255

1983	468 228
1984	505 001
1985	532 000
1986	562 001

2350

1983	433 474
1984	505 001
1985	532 000
1986	562 001

2355
1987 .... 600 000

2355N
1987 .... 600 000

2440
1976 .... 235 210
1977 .... 258 106
1978 .... 280 789
1979 .... 305 501

1980 .... 335 625
1981 .... 362 173
1982 .... 376 746

2510
1966 .... 1 000
1967 .... 8 958
1968 .... 14 291

2520
1969 .... 17 000
1970 .... 19 416
1971 .... 22 000
1972 .... 22 911
1973 .... 23 865

2550
1983 .... 433 480
1984 .... 505 001
1985 .... 532 000
1986 .... 562 001

2555
1987 .... 600 000

2630
1974 .... 188 601
1975 .... 213 360

2640
1976 .... 235 313
1977 .... 258 106
1978 .... 280 789
1979 .... 305 505
1980 .... 335 628
1981 .... 362 175
1982 .... 376 744
1983 .... 388 347

2750
1983 .... 433 494
1984 .... 505 001
1985 .... 532 000
1986 .... 562 001

2755
1987 .... 600 000

2840
1977 .... 214 909
1978 .... 264 711
1979 .... 304 654

2855N
1987 .... 600 000

2940
1980 .... 350 586
1981 .... 390 496
1982 .... 418 953

2950
1983 .... 433 508
1984 .... 505 001
1985 .... 532 000
1986 .... 562 001

2955
1987 .... 600 000

3010
1961 .... 1 000
1962 .... 10 801
1963 .... 32 400

3020
1964 .... 50 000
1965 .... 68 000
1966 .... 84 000
1967 .... 97 286
1968 .... 112 933
1969 .... 123 000
1970 .... 129 897
1971 .... 150 000
1972 .... 154 197

3150
1985 .... 532 000
1986 .... 562 001
1987 .... 587 950

4000

1969	211 422
1970	222 143
1971	250 000
1972	260 791

4010

1961	1 000
1962	20 201
1963	38 200

4020

1964	65 000
1965	91 000
1966	119 000
1967	145 660
1968	173 982
1969	201 000
1970	222 143
1971	250 000
1972	260 791

4030

1973	1 000
1974	6 700
1975	10 153
1976	13 022
1977	15 417

4040

1978	1 000
1979	3 199
1980	6 033
1981	8 707
1982	11 727

4050

1983	1 000
1984	3 501
1985	5 001
1986	6 501
1987	7 001

4230

1973	1 000
1974	13 000
1975	22 074
1976	28 957
1977	35 588

4240

1978	1 000
1979	7 434
1980	14 394
1981	20 186
1982	25 670

4250

1983	1 000
1984	6 001
1985	9 001
1986	11 001
1987	12 501

4320

1971	6 000
1972	17 031

4430

1973	1 000
1974	17 500
1975	33 050
1976	47 222
1977	62 960

4440

1978	1 000
1979	14 820
1980	29 539
1981	42 665
1982	56 346

4450

1983	1 000
1984	11 001
1985	18 001
1986	22 001
1987	24 001

4520

1969	1 000
1970	7 005

4620

1971	10 000
1972	13 692

4630

1973	1 000
1974	7 022
1975	11 717
1976	18 392
1977	25 794

4640

1978	1 000
1979	7 422
1980	13 860
1981	19 459
1982	25 729

4650

1983	1 000
1984	7 001
1985	10 001
1986	12 501
1987	14 001

4840

1978	1 000
1979	4 233
1980	7 539
1981	11 042
1982	14 933

4850

1983	1 000
1984	5 001
1985	8 001
1986	10 001
1987	11 001

5010

1963	1 000
1964	4 500
1965	8 000

5020

1966	12 000
1967	15 650
1968	20 399
1969	24 038
1970	26 624
1971	30 000
1972	30 608

6030

1972	33 000
1973	33 550
1974	34 586
1975	35 400
1976	36 014
1977	36 577

7020

1971	1 000
1972	2 006
1973	2 700
1974	3 156
1975	3 579

7520

1972	1 000
1973	1 600
1974	3 054
1975	4 945

8010

1960	1 000

8020

1964	1 000

8430

1975	1 000
1976	1 690
1977	3 962
1978	5 323

8440

1979	1 001
1980	2 266
1981	3 758
1982	5 235

8450

1982	1 000
1983	2 000
1984	3 501
1985	5 001
1986	5 501
1987	6 001

8630

1975	1 000
1976	2 382
1977	5 222
1978	7 626

8640

1979	1 500
1980	3 198
1981	5 704
1982	7 960

8650

1982	1 500
1983	3 000
1984	5 001
1985	7 001
1986	8 001
1987	8 501

8850

1982	2 000
1983	4 000
1984	5 101
1985	6 001
1986	6 501
1987	7 001

Test Numbers, years, and rated horsepower results of John Deere Tractor Tests conducted by the Nebraska testing laboratory are listed.

Please note: Prior to 1959, horsepower was reported three ways: observed horsepower; "corrected horsepower", that is, observed horsepower mathematically adjusted to reflect a standard condition (60° F and sea level barometric pressure); and rated horsepower, that is, "corrected" belt horsepower multiplied by 0.85 and "corrected" drawbar horsepower multiplied by 0.75. Nebraska Test horsepower figures used throughout this booklet are rated horsepower, figures which most closely reflect power available in average tractor working conditions.

Beginning in 1972, all tractors had diesel engines.

Test Number	Model	Rated hp Drawbar	Rated hp Belt Pulley
	1920		
1	Waterloo Boy (N)	12	25
	1924		
102	"D"	15	27
	1926		
146	"D"	15	27
	1928		
153	"GP"	10	20
	1931		
190	"GP"	15.52	24.30
	1934		
222	"A"	16.22	23.52
	1935		
232	"B"	9.28	14.25
236	"D"	24.02	37.37
	1937		
295	"G"	20.70	31.44
	1938		
305	"B"	10.76	16.86
312	"H"	9.68	12.97
313	"L"	7.01	9.27
	1939		
335	"A"	20.12	26.33
	1940		
350	"D"	30.46	38.11
	1941		
366	"B"	14.08	17.46
373	"LA"	10.46	12.93
378	"AR"	20.35	26.30
	1947		
380	"B" (Gas)	19.13	24.39
381	"B" (All Fuel)	16.64	20.68
383	"G" (All Fuel)	27.01	33.83
384	"A" (Gas)	26.48	33.53
387	"M"	14.39	18.21
	1949		
406	"R" (Diesel)	34.27	43.32
423	"MT"	14.08	18.33
429	"AR" (Gas)	26.16	33.24
	1950		
448	"MC"	13.70	18.89
	1952		
472	60	27.71	35.33
486	50	20.62	26.32

Test Number	Model	Rated hp Drawbar	Rated hp Belt Pulley
	1953		
490	60 (All Fuel)	22.57	28.27
493	70	33.16	42.80
503	40	17.16	21.45
504	40S	16.77	21.13
505	40C	15.11	21.24
506	70 (All Fuel)	30.75	38.22
507	50 (All Fuel)	17.42	21.89
513	60 (LP)	28.59	35.90
514	70 (LP)	34.58	44.17
	1954		
528	70 (Diesel)	34.25	43.77
	1955		
540	50 (LP)	21.90	27.45
546	40S (All Fuel)	14.25	17.76
567	80 (Diesel)	46.32	57.49
	1956		
590	520 (LP)	25.63	32.38
591	620 (LP)	34.34	42.79
592	520 (All Fuel)	18.58	22.62
593	720 (LP)	40.63	50.67
594	720 (Diesel)	40.25	50.01
597	520 (Gas)	25.73	32.79
598	620 (Gas)	33.12	41.38
599	420-W	20.31	24.83
600	420S (All Fuel)	16.42	19.95
601	420C	18.09	25.26
604	620 (All Fuel)	24.50	30.33
605	720 (Gas)	39.79	50.25
606	720 (All Fuel)	30.97	38.53
	1957		
632	820 (Diesel)	52.25	64.26

Test Number	Model	Maximum PTO hp
	1959	
716	435 (Diesel)	32.91
	1960	
759	4010 (Gas)	80.96
760	4010 (LP)	80.60
761	4010 (Diesel)	84.00
762	3010 (Diesel)	59.44
763	3010 (Gas)	55.09
764	3010 (LP)	55.39
	1961	
798	1010C (Diesel)	36.31
799	2010RU (Diesel)	46.67
800	2010RU (Gas)	46.86
801	1010C (Gas)	35.93
802	1010RU (Gas)	36.13
803	1010RU (Diesel)	35.99
	1962	
828	5010 (Diesel)	121.12
829	2010C (Gas)	47.45
830	2010C (Diesel)	47.72
	1963	
848	3020 (Diesel)	65.28
849	4020 (Diesel)	91.17
850	4020 (Gas)	88.09
851	3020 (Gas)	64.14
852	3020 (LP)	64.70
853	4020 (LP)	90.48
	1965	
913	2510 Power Shift (Gas)	49.57
914	2510 Syncro-Range (Gas)	53.74
915	2510 Power Shift (Diesel)	50.66
916	2510 Syncro-Range (Diesel)	54.96
930	4020 Syncro-Range (Diesel)	94.88
	1966	
934	4020 Syncro-Range (LP)	94.57
935	1020 (Gas)	38.82
936	2020 (Gas)	53.91
937	1020 (Diesel)	38.92
938	2020 (Diesel)	54.09
939	4020 Syncro-Range (Gas)	95.59
940	3020 Syncro-Range (Diesel)	71.26
941	3020 Syncro-Range (Gas)	70.59
942	3020 Syncro-Range (LP)	70.66
947	5020 (Diesel)	133.25
	1968	
991	1520 (Diesel)	46.52
992	2520 Syncro-Range (Diesel)	61.29
993	2520 Power Shift (Diesel)	56.28
	1969	
1002	2520 Power Shift (Gas)	56.98
1003	2520 Syncro-Range (Gas)	60.16
1004	1520 (Gas)	47.86
1010	3020 Power Shift (Gas)	67.13
1011	3020 Syncro-Range (Gas)	71.37
1012	4020 Power Shift (Gas)	95.66
1013	4020 Syncro-Range (Gas)	96.66
1014	4520 Power Shift (Diesel)	122.36
1015	4520 Syncro-Range (Diesel)	123.39
1023	4000 (Diesel)	96.89
1024	4020 Power Shift (Diesel)	95.83
1025	5020 (Diesel)	141.34
	1970	
1050	4320 Syncro-Range (Diesel)	116.55
	1971	
1063	7020 (Diesel)	146.17
1064	4620 Power Shift (Diesel)	135.62
1073	4620 Syncro-Range (Diesel)	135.76
1084	2030 (Gas)	60.34
1085	2030 (Diesel)	60.65
	1972	
1100	6030	175.99
1101	7520	175.82
1110	4430 Quad-Range	125.88
1111	4030 Quad-Range	80.33
1112	4230 Quad-Range	100.32
1113	4630 Power Shift	150.66
	1973	
1146	830	35.30
1147	1530	45.38
	1974	
1157	2630	70.37
	1975	
1179	8430	178.16
1180	8630	225.59
1191	2040 (164 cu. in. engine)	40.86
1192	2240	50.37
	1977	
1249	2840	80.65
1263	4840	180.63
1264	4640 Quad-Range	156.30
1265	4440 Quad-Range	130.58
1266	4240 Quad-Range	110.94
1267	4040 Quad-Range	90.80
	1978	
1280	850	22.27
1281	950	27.36
	1979	
1323	8440	179.83
1324	8640	228.75
1327	1050	33.41
	1980	
1350	2040 Collar-Shift (179 cu. in. engine)	41.25
1351	2940	81.17
1357	4440 Power Shift	130.41
1358	4640 Power Shift	155.96
1359	4040 Syncro-Range	90.31
1360	4040 Power Shift	90.79
1361	4240 Syncro-Range	110.81
1362	4240 Power Shift	111.06
	1981	
1393	2040 TSS	40.44
1394	2240 TSS	50.90
	1982	
1434	8850	303.99
1435	8650	238.56
1436	8450	186.98
1437	1250	40.70
1457	4050 Power Shift	100.95
1458	4250 Power Shift	120.86
1459	4450 Power Shift	140.43
1460	4650 Power Shift	165.52
1461	4850 Power Shift	192.99
	1983	
1469	2150	46.47
1470	2350	56.18
1471	2550	65.94
1472	2750	75.35
1473	2950	85.37
1474	4050 Quad-Range	101.49
1475	4250 Quad-Range	120.21
1476	4450 Quad-Range	140.33
1477	4650 Quad-Range	165.70
1504	750	18.54
1505	1450	51.39
1506	1650	62.22
	1986	
1589	3150	96.06
1590	4050 Quad-Range (359 cu. in. turbo engine)	105.89
1591	4050 Power Shift (359 cu. in. turbo engine)	105.69
1605	2755	76.57
1606	2955	86.22